Algebraic 3-D Modeling

Algebraic 3-D Modeling

Andreas Hartwig
Dresden University of Technology
Department of Mathematics
Institute of Geometry
Dresden, Germany

A K Peters
Wellesley, Massachusetts

Editorial, Sales, and Customer Service Office

A K Peters, Ltd.
289 Linden Street
Wellesley, MA 02181

Library of Congress Cataloging-in-Publication Data

Hartwig, Andreas, 1949–
Algebraic 3-D modeling / Andreas Hartwig.
p. cm.
Includes bibliographical references and index.
ISBN 1-56881-023-7
1. Digital computer simulation. 2. Computer graphics. I. Title.
QA76.9.C65H38 1996
003'.3—dc20 96-15597
CIP

Printed in the United States of America
00 99 98 97 96 10 9 8 7 6 5 4 3 2 1

Contents

Preface

This book is written for computer scientists, mathematicians and design engineers, including students who are interested in modeling geometric objects in three-dimensional space. Based on practical experience, we have developed a concept that explains mathematical relationships between different types and implementations of modeling systems. This concept will help the reader to understand and implement modeling software for different applications.

It is often difficult to establish a mathematically based comparison of the many tools that exist for generating geometric objects and to recognize the practical limitations of comparable software. Pricing and hardware requirements further complicate such a comparison. The integration of one's own modeling subsystems into existing systems in order to handle special applications may lead to restrictions in terms of program size, stability, time complexity, and so on. It is therefore desirable to list the practical problems and difficulties in describing the generation and application of three-dimensional geometric models on a computer without discussing all existing and possible software in detail.

To achieve this in a way that allows a fair comparison between existing systems, we use mathematical concepts in an algebraic context. This allows us to integrate the various approaches based on different mathematical theories, such as topology, algebra, combinatorics, etc. which are part of present or currently developed modeling systems. Connections between the different mathematical theories reflect the limits of the various modeling systems. We do not claim an all encompassing algebraic treatment of 3-D modeling. We concentrate on those situations that arise from the simultaneous use of half-space and boundary representations. In this case it is possible to compare algebraic and geometric parameters of the desired representations. Splines are not considered in this context because in many problems using splines, the algebraic and geometric parameters are one and the same. Furthermore, algebraically oriented algorithms are not the object of our discussion, although representations are compared in an algebraic sense. The small language modeler was added in order to explain problems of 3-D modeling in a practical way. It shows the possibilities of parametric modeling.

I hope the book will help the reader to better understand the mathematical aspects of modeling and to appreciate modeling as a more automatic process than it has been seen until now.

The demo modeler GECO can be downloaded by ftp from
ftp: tu-dresden, path: pub/soft/msdos/3dmod/geco

The author can be reached at hartwig@tudurz.urz.tu-dresden.de

Acknowledgments

I would like to thank Wolfgang Straßer who motivated me to write down and to continue work on the topic of this book. I wrote most of the book while at Straßer's institute in Tübingen.

I especially thank Duong Kieu Hoa for her theoretical and practical work which led to the foundational ideas of this text. She spent much time discussing the relationship between algebraic and topological properties, and she worked out the first ideas of generalized algebras.

I thank Thomas Nestler as well. He described and discussed the morphism between the Boolean algebra and a special sign algebra that is able to classify geometric properties like edges and faces of algebraically described polyhedra to produce boundary representations. Furthermore, I would like to thank Reinhard Klein, Thomas Jerowsky, Klaus Mager, Wolfgang Burmeister, Anja Pruchnewski, and Gabriele Reinsberg for improving the contents of the book.

The small modeler introduced in this book was often used in Dresden on small computers to teach modeling of geometric objects and to illustrate practical problems which arise in modeling. I thank András Hernadi, Karla Nestler, Helga Mettke, Gerd Zimmermann, and Joachim Wolf for contributing essential parts to the program. For testing, teaching, and applying the modeler, I thank Klaus Jacob, Klaus-Dieter Hanisch, Holger Lieberwirth, Ines Gubsch, Heinz Eberl, Heidemarie Engelhard, Christiane Kamusella, Harald Kogel, Gerd Richter and René Marschner.

The text was written concurrently with my lectures which were attended by both mathematicians and engineers interested in computer science. In each of the related "proseminars," one or two of the participating students introduced and prepared a single chapter for discussion—my thanks to them for their comments.

I also would like to thank my family for their understanding and support of my work.

1

Motivation

1.1. The Objects of Modeling and Their Representations

Geometric objects will be represented as point sets in three-dimensional Euclidean space E^3. In an actual implementation, the following applications of the sets should be supported. It should be possible, for example, to change the location of the sets in space, to point-out inclusionary relations and to estimate distances, contents of surfaces and volumes of the represented objects. Moreover, it should be possible to produce cross sections and pictures on a screen or by a plotter.

In [Re 80] representations of geometric objects are discussed and it is illustrated that the CSG-representation (constructive solid geometry) and the boundary representation are the most important ones.

The *CSG-representation* of a geometric object consists of elementary sets and linking instructions according to set operations. The elementary sets have typical geometric forms like half-spaces, triangles, cubes, or cylinders.

The *boundary representation* of a set includes the description of its oriented boundary parts. The orientation characterizes the location of the inner points of the objects.

While the CSG-representation uses the algebraic way for set modeling, the boundary representation focuses on the topological characteristics of

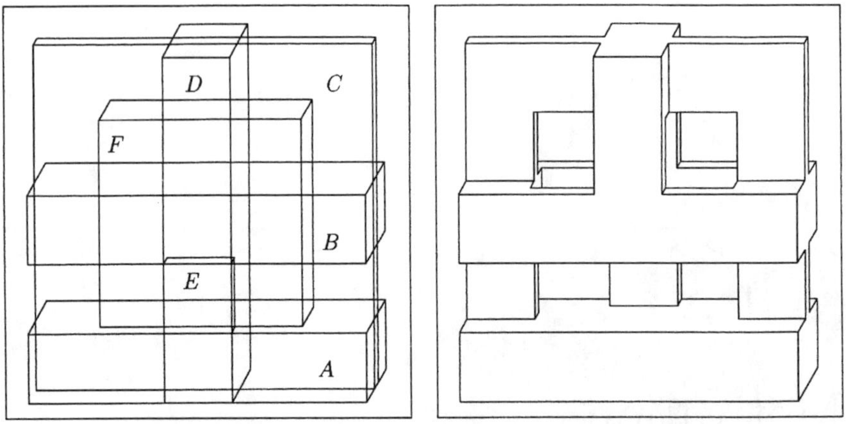

$H = (A \cup B \cup C \cup D) \setminus (E \cup F)$ H

Figure 1.1. Modeling of a boundary representation of a set H as a Boolean set operation of six rectangular solids.

the result. Both approaches have advantages, and we will try to utilize both.

Because of the finite representation of numbers on the computer, a desired object has to be approximated in most cases. So, in the following we will define a CSG-representation or a boundary representation of approximated sets in E^3 as a *computer-internal description.*

1.2. Modeling with Boolean Set Operations

When describing objects by sets, set operations correspond to the steps needed to model the objects. Therefore, the Boolean set operations enter the modeling software (see Figure 1.1).

The left hand side of Figure 1.1 shows the boundary representation of elementary objects which, together with the given set formula, constitute a CSG-representation of H. The right hand side is a picture of the boundary representation of H.

In a CSG-representation, a set operation is expressed only by adding the set formula to the operands. In our approach to describe algebraic operations with sets, we will not demand that the operations are necessarily Boolean ones. Therefore, CSG-representations are a simple way to represent other set operations.

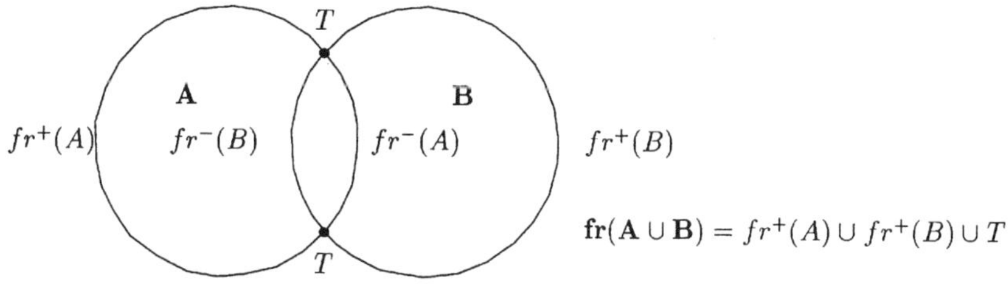

Scheme 1.2. Linking the boundary of $A \cup B$ by corresponding parts.

By linking the sets, topological issues are changed. They are expressed by the boundary representation of the set operation result, which can only be obtained by complicated computations. The basic idea of algorithms for linking the boundary representations by means of Boolean set operations consists of the following:

- decomposing the boundaries of operands into parts which are placed completely in either the *interior* (int), *frontier* (fr) or *exterior* (ex) of the result,
- classifying the location of these parts with respect to the operands by means of incidence tests and
- composing the ascertained frame parts of the results into the desired representation.

Scheme 1.2 of a binary Boolean set union $A \cup B$ of the sets A and B illustrates this procedure.

Here, the symbols $fr^{-}(A)$, $fr^{+}(A)$, $fr^{-}(B)$ and $fr^{+}(B)$ denote corresponding boundary parts of the operands and T signifies the intersection of the frontiers $fr(A)$ and $fr(B)$.

1.3. Stability of Boolean Set Operations

In a mathematical model, it is desirable to algebraically close the set operations with respect to the related class of operands. That means that all results from set operations may be used as operands again. This ability will be achieved in the mathematical model by using an algebra as a basis.

Putting the boundary representation into a computer and calculating new representations, errors are produced by limited precision arithmetic, which will have serious consequences on the topological properties of the

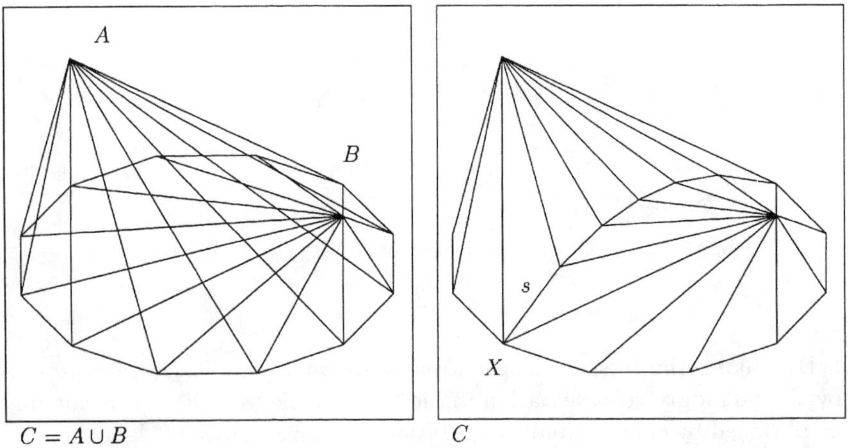

Figure 1.3. The Boolean set operation of two cones can contain a numerically unstable point X.

result. So, algebraic closure of the internal representation cannot be guaranteed in general. The following three factors are known to be detrimental to Boolean operations:

- the approximation precision of the computer-internal description of the operands,
- the stability of algorithms that determine the boundary separation and
- unstable incidence tests which produce conflicting topological properties of the computer-internal description.

Problems stemming from the first factor can be avoided during computation by insuring algebraic closure in the approximation class. There are suitable possibilities, which are not explained here in detail.

Problems stemming from the second factor cannot be avoided as easily. For example, difficulties may arise during the computation of the intersection between the boundaries of two operands. Figure 1.3 shows a numerically unstable point X, which appears in the union of two different cones with the same base circle. An intersection algorithm for finding the intersection line s improves its stability away from X.

In the corresponding non-approximated model, this point X is described only by the base plane, a tangential plane and by a limit of a sequence of better conditioned points of s. Therefore, simple intersection algorithms are not sufficient to produce the intersection. It is necessary to consider

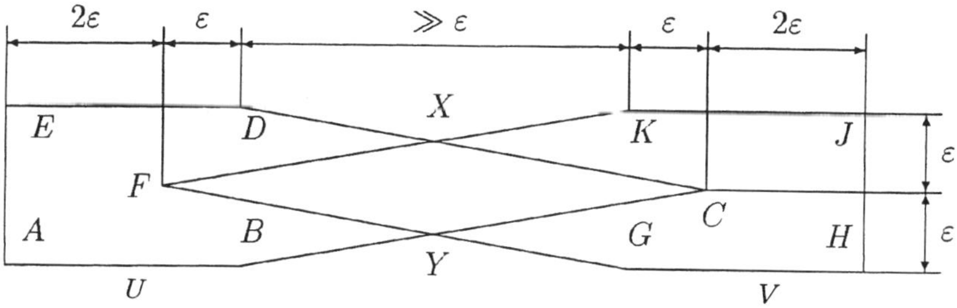

Figure 1.4. An unfavorable example for computing the boundary representation of the difference of two planes U and V that is dependent on the object precision and the distance t.

the representation model in a new sense. Our aim should be to provide better conditioned separating sets and a new description of the desired intersection.

The third factor, the loss of algebraic closure, manifests itself in the fact that computed boundary parts do not allow composition of a common boundary representation. For an illustration, let us consider two pentagons U and V with vertices $ABCDE$ and $FGHJK$ in Figure 1.4. Let the positive real number ε be the object precision of the internal description and let t be a real number which is significantly larger than ε. The intersection points X and Y cannot be used for boundary separation. It is necessary to compute the points X and Y with significantly higher precision than ε, assuming a sufficiently large distance t. In case the points X and Y are computed by ε then a correct connection between boundary parts does not exist.

We are now ready to get a useful boundary representation of the difference $U \setminus V$ according to ε-precision, namely one which is only described by a closed directed-edge sequence of line segments and ε-connections. In the ε-sense, the point D belongs to $U \setminus V$, but C does not. Therefore, it cannot be decided whether the line segment CD can be added to the result or not. Moreover, other vertices are farther away from D than ε. It is far from obvious how to arrive at a manageable representation.

In addition to these problems, other difficulties arise: suppose we want to define an object not far from the lower or upper bound of the object precision of the number representation. It is possible to construct examples which break the Boolean rules like associativity or commutativity of the union. In the following chapters we will discover such approximations of the operands and results of set connections. We will also classify their rules such that algebraic closure of the approximation class will be possible in the computer-internal description.

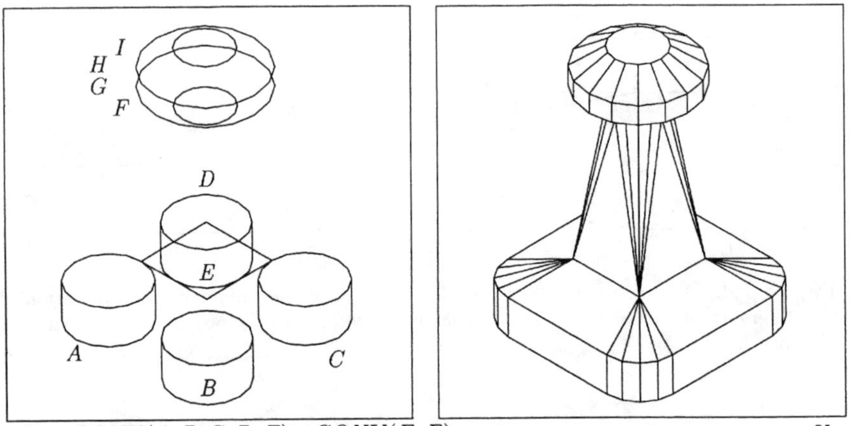

$K = CONV(A, B, C, D, E) \cup CONV(E, F)$ $\qquad$ K

$\cup\ CONV(F, G, H, I)$

Figure 1.5. Union of three convex hulls ($CONV$) into one object.

1.4. Looking for Algebraic Generalization

For the description of complicated objects, set operations other than the Boolean ones have proved to be efficient in the sense of linguistic formulation and of stable computation.

Hull Constructions

Forming several elementary objects into the convex hull is particularly attractive (see Figure 1.5). In this figure, an object is formed by three convex hulls. The first is constructed by four cylinders A, B, C and D, and the rectangle E; the second is constructed by E and a circle F; and the third is constructed by the four circles F, G, H and I.

The resulting object, composed of the elementary objects $A, B, \cdots, I$, cannot be created by means of Boolean set operations. However, intersection algorithms are not necessary to produce the boundary representation of the result. Instead of using intersection algorithms, tangential planes can be found. The advantages in this approach motivates us to look for a more general hull concept including the closure operator of Kurakowski's closure axioms [Ku 68].

Generalization of Algebras

It is interesting to obtain a more abstract model than the Boolean set algebra. Namely, one which contains the Boolean set operations as well as the

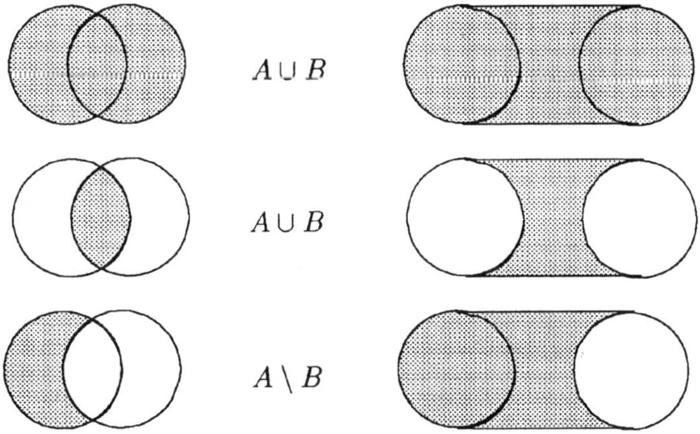

Scheme 1.6. Modification of union (∪), intersection (∩) and difference (\).

convex hull, more general closure operations and other modeling operations by special specifications. The basic idea will be to use a third operand in the binary Boolean set operations which has to be chosen dependent on the other two operands. In this way we can represent other binary operations, and so we are interested in the rules in general and for each specification.

Scheme 1.6 shows a simple modification of the Boolean union (∪), intersection (∩) and difference (\) between two operands A and B into corresponding hull operations.

Such operations deal with connections between two given geometric objects. We will define this modification later by means of a reduced convexity definition. The new algebra will require answers to some questions, including the following list.

- What is the relationship between the Boolean and new algebra?
- Are the operations closed?
- What are the simplest generators of the operation?
- How many possible binary operations are there?
- How can we describe the new boundary separation and connection?
- What abstract model can describe the Boolean and the new algebras together?

1.5. The Aim

This text deals with difficulties of geometric modeling. The available modeling software for representing three-dimensional geometric objects, although commonly used, has many limit cases with non-desirable effects. They disturb the direct application of the modeling software as well as the application of the modeling software embedded in a suitable environment (graphic interaction, data base, etc.). Here we only concentrate on a direct application of the modeling software and problems that are related to the computer internal description of geometric objects, thus neglecting the fact that problems are reduced in CAD systems by using a suitable environment.

In order to capture these special problems, we have to consider the entire complexity of the modeling structure. Therefore we first translate the following computer science oriented notions into mathematical ones.

computer science	**mathematics**
model	tuple (A, F) A ... base set F ... set of functions on A
operationally closed model	algebra $\mathcal{A}(A, F)$
object - elementary object - functional object - hybrid object	 $A_i \in A$ $F_j \in F$ tuple (A_i, F_j)
representation	algebra $\mathcal{R}(A', F')$ as a map of $\mathcal{A}(A, F)$
operationally suitable representation	morphism between $\mathcal{A}$ and $\mathcal{B}$

On this level, we do not understand the notion of the representation as a notion for describing computer internal objects and operations. This is the aim of the *data structure*. It consists of a set D of data and a set P of programs. A *consistent data structure* is a suitable representation of the closed model with $A' = D$ and $F' = P$.

Other representations of a model are geometric, numeric, algebraic structures and so on. Those are called *model specifications*, and they are distinguished in the different way of considering the modeler. So, different mathematical theories have to be taken into consideration together with special directly or indirectly used notions describing the modeling process. The specifications can be departed by the *object structure* of a modeler which consists of *elementary structure*, *functional structure* and *hybrid structure*.

object structure	**model specification**	**theory reference**	**example object**
elementary	topolologic geometric numeric	ball shape number	regularized set polyhedron real number
functional	combinatoric algebraic	complexity operation	separating test set operation
hybrid	data language	architecture communication	CSG and boundary modeling language

The mentioned problems arise mostly between two such model specifications. If a morphism between specifications exists then they are called *suitable model specifications.* Now, we can formulate our ideal in this context.

> *We want to get a so called* **correct modeler** *which uses only suitable model specifications.*

Therefore, we will examine descriptions using suitable model specifications, hoping that we can expend our assessment about any geometric modeling software. The title of this book was chosen because our aim of suitable model specifications is algebraically motivated. At the same time we search for a uniform theoretical concept of modeling that improves the understanding as well as the implementation of software describing three-dimensional geometric objects.

1.6. Remarks on the Chapters

The individual chapters are independently readable. Their contents will not globally be discussed in the chapters themselves. Therefore we want to motivate the readers of the following chapters by giving an overview of the book's contents. The notations in parenthesis will be defined later.

Polyhedra (Chapter 2)

Modeling will be described as the mathematical model of polyhedra. For the description of geometric objects in three-dimensional space, two ways, based on two different global mathematical theories, are possible. Topology describes those objects by a union of a finite number of sets of varying dimensions (topologic varieties, the manifolds). The boundary representation as an internal description of polyhedra prefers this concept.

On the other hand, the CSG-representation is often based on systems of inequalities. Therefore it is necessary to define polyhedra by means of algebraic expressions and equations, too. This algebraic way needs the handling of systems of equations for defining the polyhedra as solution sets (algebraic varieties). By means of equations, it is also possible to give a mathematical definition of the intuitively well-understood notions of polyhedron edges and vertices. Its logical (Boolean) characterization shows a practical test to decide whether a curve piece is an edge of the desired polyhedron or not.

Both ways have to be combined to consider the CSG-representation as well as the boundary representation in one common concept, and they will be combined in forms of hierarchical polyhedron notions (simple polyhedron, i-polyhedron, general polyhedron). In some cases it is necessary to operate only in one of the hierarchical levels (regularizing set operations with polyhedra). Furthermore, operations with sets of different dimensions (general polyhedra, assemblies) are interesting. A special, topologically motivated representation (dimension representation), introduced in this context, allows us to compute set operations with general polyhedra by means of regularizing operations in each of its components.

Boundary Models (Chapter 3)

Face or volume models of geometric objects demand of its boundary representations that it is always possible to decide whether any given point belongs to an object or not (incidence test, complete structure). Data structures often differ in the importance of their geometric references (explicit and implicit structures). As a rule, the kind of data structures characterizes the kind of algorithms carrying out geometric operations. Therefore in the special case of linearly restricted polyhedra, implicit structures of boundary representations and their transformations inside one another are motivated.

Modelers produce large amounts of data. The detail of difficult objects are put-out on a screen by the size of the screen pixel. Another way to improve the object geometrically will be introduced by an example (of an excavator); hull constructions help to build geometric abstracts without detail in its special geometric kind.

The combinatorial aspect of data structures is determined by the question whether two given geometric objects are disjoint or not. The most widely used algorithms that give a sufficient answer utilize tests in each of the two or three single vertex components. Those box tests are not able to confirm the disjointedness in all cases. Sharp separations are desired. Especially, if we want to transfer CSG into boundary representation, then we have no points that we can use in a box test. Therefore the simplex method of linear optimization, extended by a special characteristic quotient (second characteristic quotient), will be introduced to produce sharp separations and also to enumerate the vertices of each boundary piece. The back transformation of boundary into CSG-representation is possible in different ways, but it is of lesser importance.

The Small Language Modeler (Chapter 4)

The GECO modeler is added to show some possibilities as well as difficulties in practice. Internally it uses a transformation algorithm CSG-representation (half-space model) in boundary representation (surface structure).

The elementary and convex objects are stored in the boundary representation. They are transferred into conjunctions of half-spaces to get a half-space model for computing the boundary representation of a Boolean set formula over convex objects. It would not be necessary to go the roundabout way with the half-space model, this shows the experimental character of the modeler.

Boolean set operations are available only in a pure dimension, second or third, but tree operations build compound objects. The convex hull operation expands the possibilities of connections. The produced objects can be described in different ways (hybrid model of CSG- and boundary representation, the boundary as surface, edge or vertex structure).

The software is given as a pure modeling module without a user interface. The available geometric language to describe three-dimensional objects by words permits a short documentation of the constructed object on paper as well as a parametrization of geometric constructions.

GECO tries to keep track of the modeling process. It is well-known that a high level of graphics interactivity puts the designer under stress. Another motivation for describing the desired object by language and not by graphical interactions is the size of the models. The screen is obviously not able to show large models in each case while their detail is being modeled.

The following chapters should show the limits of modelers like GECO and the limits of the stability of Boolean set operations that are used together with hull operations. We also address the question of other algebraically motivated set operations with more generality. The aim is a fair evaluation of the different kinds of modeling operations in the sense of algebraically closed operations in order to get clear mathematical models.

Algebraic Model (Chapter 5)

Considering only classes of geometric objects and their representations, we neglect that operations could go out of these desired classes. Theoretically it is important at each time to know the validity of the object with respect to a special class. Thus we are not forced to apply validity tests of operation results.

Set algebras are primarily Boolean algebras (see Axioms). But a lot of modeling operations cannot be described only in terms of the Boolean operations. Therefore this book suggests to weaken the Boolean Axioms depending on a given Boolean algebra. So it will be possible to find a new algebraic model (natural modeling algebra) that retains those parts of the Boolean algebras that are necessary for the original modeling idea (Euler Venn diagrams) with set operations and some adequate terms (weak disjunctive normal form and basic operations) to declare binary operations and for representing all of the possible operations succinctly.

Computation of Algebraic Manifolds (Chapter 6)

Firstly the chapter deals with the question of how the actual coordinates of the vertices are to be rounded, and it motivates an equidistant point grid to make translations by a corresponding grid vector without rounding (discretization of the Euclidean space E^3). After that, special domains (restricted and doubly restricted domain) are introduced for correct computation of scalar products. All of the interesting operations in the E^3, e.g., vector products, will be based on scalar products.

Based upon the correctness of scalar products over an equidistant point grid, we can consider the precision of polynomial equations describing geometric forms (form and shape). In that way the polynomials are split into scalar products. Exact arithmetic equations defined by a number of grid points (geometric parameters of the equation) need a large number of digits (algebraic parameters of the equation). Rounding is necessary up to now.

An intuitive way to operate with rounded algebraic parameters is introduced by examples and by a special definition (geometrically suitable) that reflects a relation between geometric and algebraic parameters in the sense of ε-precision for describing equations with operators such as scalar products. By means of this notion it is possible to define a rounding of the algebraic parameters such that the geometric form of the rounded objects differs only by an ε-distance.

The precision of intersection curves is important. The intersection of two implicitly defined surfaces of second degree can be described by Cardano's formula. Zeros of polynomial equations of degree four can only be computed by writing out explicitly the single space coordinates depending on a curve parameter. Therefore we estimate the coefficients of explicit curves by usual notions (relative and absolute condition). Later we will see how many digits are necessary to describe models without any topological contradiction during modeling.

Topological Specification (Chapter 7)

It is not trivial to distinguish the model parts of different dimensions (regular set algebras). Therefore it is necessary to explain the set operations in each dimension as well as the operations with general polyhedra in forms of the assemblies mentioned above. The mapping (morphism) of the set operations between the desired set algebra and the regularization components is of algebraic interest in order to construct pure and therefore intuitive modeling operations.

The point grid permits some definitions of regularizing. For example, it is conceivable, that two-dimensional parts of an assembly model have to be approximated by one-dimensional objects.

Another difficulty to build modeling set operations arises from topological discontinuities that reflect only ε-small tolerance of the algebraic parameters. Such topological discontinuities lead to unstable incidence tests in set operations. They can be avoided in two ways:

- We may use an algebraic-topologically suitable representation.
- Or it is possible to rid the operation model of the discontinuities.

The first way needs the discussed number of digits that is only advantageous in linearly restricted models. Then the usual arithmetic processors compute products large enough. In the quadratic case we need about a hundred decimals to make correct incidence tests. The second way sometimes leads to correct but unwanted effects in some objects. Nevertheless, they are understandable and can be produced in each case by specific algorithms.

Regularization as well as this topological modification lead to the algebraic concept of natural modeling algebras.

Modeling Principle (Chapter 8)

The modeling principle perceives modeling as the union of several different ways to introduce one particularly desirable set algebra on a computer, and it combines possible modeling operations in one and the same algebraic concept.

After some prerequisite notions (separation and connection sets), the modeling elements describe modeling parts (of an algebraic, topologic and constructive type) that belong to different fundamental theory parts, including the operations with assemblies by means of the dimensions representation and the regular set algebras.

Constructive specifications of the natural modeling algebra classify well-known modeling operation as well as new ones (for example special hull algebras).

2

Polyhedra

Geometric objects have special topological properties which have to be taken into consideration. A proliferation of topologically interesting point sets should be avoided in order to describe technically relevant objects: they complicate the data structures of the model. Therefore, we confine ourselves to representation by means of polyhedra.

2.1. Topological and Algebraic Variety

Let X be a point of n-dimensional Euclidean point space E^n. We denote the coordinate vector assigned to X by $x = (x_1, ..., x_n) \in R^n$ with reference to the Cartesian coordinate system of E^n. In the following, we will be interested in Euclidean space E^3 and its lower dimensional subspaces. The notion of a variety is immediately related to the dimension of the space. It is possible to motivate this notion topologically as well as algebraically.

Topological Variety

In the topological sense, the notion of variety is defined by a so-called homeomorphic mapping. It is sufficient to define this mapping by subsets of E^3 only. Let A and B be two subsets of E^3 and let f be a bijective mapping of A to $B = f(A)$. This mapping f is called a *homeomorphic mapping* if and only if f and its inverse mapping f^{-1} are continuous. The set B is called a *homeomorphic image* of the set A.

A ball $U(X)$ denotes the set of all points in E^3 having a distance less than a sufficiently small number *eps* from the point X. In this case, we measure the distance by the Euclidean norm. A set $M \subset E^3$ is *locally homeomorphic* to a subspace E^n $(0 \leq n \leq 3)$ if each ball $U(X) \subset M$ of any point $X \in M$ is homeomorphic to E^n. Such a set M is called an *n-manifold.* The manifolds can be classified by their dimensions corresponding to their geometric application to *points* $(n = 0)$, *curves* $(n = 1)$, *surfaces* $(n = 2)$ and *solids* $(n = 3)$. The n-manifolds are identical to n-dimensional varieties in the topological sense.

Algebraic Variety

The algebraic formulation of the variety notion results from m given equations in a Cartesian coordinate system and reflects properties of their solution set. It is useful to write the m equations either in an explicit parametric or in implicit parametric form. For our purpose, we define the algebraic variety notion in an implicit form. Let $g_1(x), \cdots, g_m(x)$ be functions of an open connected set $U \subset R^3$ to R that have continuous partial derivatives; their associated solution set

$$W = \{x \in U \mid g_l(x) = 0, l = 1, ..., m\}$$

is not empty and their Jacobian

$$J = \begin{pmatrix} \frac{\delta g_1}{\delta x_1} & \frac{\delta g_1}{\delta x_2} & \frac{\delta g_1}{\delta x_3} \\ \vdots & \ddots & \vdots \\ \frac{\delta g_m}{\delta x_1} & \frac{\delta g_m}{\delta x_2} & \frac{\delta g_m}{\delta x_3} \end{pmatrix}$$

has maximal degree $3 - d$ for all points $X \in W$. Then W is called a d-dimensional variety in the algebraic sense. This definition can be explained geometrically. The i-th row vector of the Jacobi matrix is the gradient of the i-th function at location x. The solution set belonging to the i-th equation of W is a surface in three dimensional space if it is not empty. Then the gradient of this surface at a location $x = x_0$ is of interest because it satisfies all m given equations, i.e., the i-th equation defines a common point of all m surfaces in three-dimensional space. Considering the tangential equations and their normal vectors, the degree of the Jacobi matrix demonstrates the kind of the linear independence of these normals.

It is worth noting that a d-dimensional (algebraic) variety contains a d-manifold and possibly lower dimensional manifolds as well. It is not clear how many pairwise disjunct manifolds can belong to a d-dimensional (algebraic) variety. The following constraint for the functions $g_1(x), ..., g_m(x)$ is to be assumed:

(U1) The solution set W can be represented as a union of a finite number of connected point sets.

This premise (U1) has the practical consequence that only a finite number of objects can be managed on a computer. It is not an essential restriction for our purpose and even simplifies the theoretical description of the modeling objects in three-dimensional space.

Because of its definition by equations, an n-dimensional variety in the algebraic sense is called an *algebraic point* ($d = 0$), an *algebraic curve* ($d = 1$), an *algebraic surface* ($d = 2$) and an *algebraic solid* ($d = 3$). In this context, we only speak of points, curves, surfaces and solids, if no confusion can arise. As mentioned above, a d-dimensional algebraic variety consists of dimensional topological varieties of various dimensions (manifolds) and can be represented as a finite number of

$d = 3:$	solids,	surfaces,	curves	and	points,
$d = 2:$		surfaces,	curves	and	points,
$d = 1:$			curves	and	points,
$d = 1:$					points.

2.2. Supporting Sets and Regularization of Sets

Now we prepare an extension of the algebraic variety in order to describe the solution set of systems of inequalities in R^3 algebraically and topologically. These systems of inequalities and their representation in E^3 will be the models of geometric objects which are desired.

Infinitely often differentiable functions are called *analytic functions.* In the definition of an algebraic variety, we use a system G_L of a finite number of analytic functions $g_l \ \ (l \in L)$, which map R^3 into R^1. F_l denotes the solution set of the equation $g_l = 0 \ \ (l \in L)$. Let $G_L = \{g_l | l \in L\}$ be well-defined so that each intersection of any associated algebraic variety meets the constraint (U1) or is equal to the empty set. Then each intersection can be represented by a finite number of manifolds as well.

For all $i, j, k \in L$ with $F_i \neq F_j, F_i \neq F_k$ and $F_j \neq F_k$, we define the sets

$$\begin{aligned} T &= E^3 \\ T_i &= F_i \\ T_{ij} &= F_i \cap F_j \\ T_{ijk} &= F_i \cap F_j \cap F_k. \end{aligned}$$

Lemma 2.1. *The sets* T, T_i, T_{ij}, T_{ijk} $i, j, k \in L$ *are algebraic varieties of dimension at most* 3, 2, 1 *and* 0, *respectively.*

Proof. According to the constraint on the function set G_L, all functions $g_l \in G_L$ $(l \in L)$ can be written as a Taylor series

$$g_l(x, y, z) = \sum_{i,j,k=0}^{\infty} a_{ijk}^{(l)}(x - x_0)^i (y - y_0)^j (z - z_0)^k$$

at the point $P = (x_0, y_0, z_0)$. Let us consider two such functions $g_r, g_s \in G_L$ that fulfill $g_r \neq g_s$ and also

$$g_r(x_0, y_0, z_0) = g_s(x_0, y_0, z_0) = 0.$$

Inserting the point $P = (x_0, y_0, z_0)$ into the series it follows

$$a_{000}^{(r)} = a_{000}^{(s)} = 0.$$

The normals $grad \;\; g_r(x_0, y_0, z_0)$ and $grad \;\; g_s(x_0, y_0, z_0)$ of the surfaces F_r, F_s at the point P can be generated by taking derivatives of the Taylor series. They are equal to each other if and only if their corresponding coefficients are equal. Because of $g_r \neq g_s$ and $a_{000}^{(r)} = a_{000}^{(s)}$ it follows that one of the non-vanishing coefficient pairs is not equal,i.e., that the gradients of the two surfaces F_r and F_s do not have the same direction at the point $P = (x_0, y_0, z_0)$ and that T_{rs} has dimension of the algebraic variety at most 1. Analogously, the set T_{rst} $(r, s, t \in L)$ has dimension at 0. It is trivial that T_r $(r \in L)$ has dimension at most 2 and T is three-dimensional. □

Definition 2.2. *Let* $i, j, k \in L$ *and let the algebraic dimensions of* T, T_i, T_{ij} *and* $Tijk$ *be* 3, 2, 1 *and* 0, *respectively. Then we call*

T	*supporting space,*
T_i	*supporting surface,*
T_{ij}	*supporting curve and*
T_{ijk}	*supporting point.*

A set $T_* \in \{T, T_i, T_{ij}, T_{ijk}\}$ *is only called* **supporting set**. *We denote a supporting set of dimension* n *by* n*-set.*

Using these definitions, we now modify a ball $U(X)$ of a point $X \in E^3$ relative to T_* by

$$U(X) = U(X) \cap T_*.$$

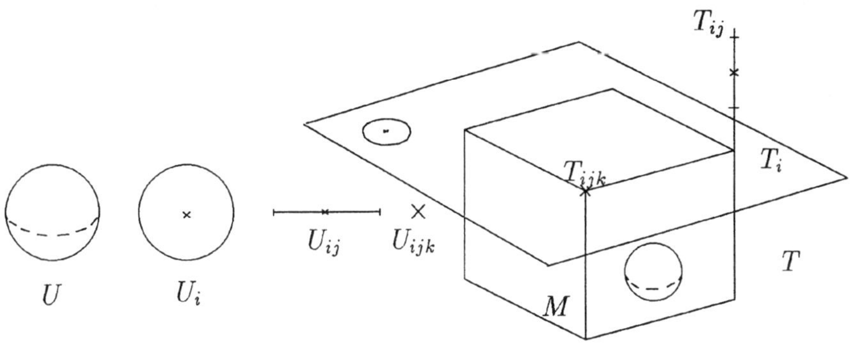

Figure 2.1. Schematic representation of supporting sets and relative balls of different type.

Figure 2.1 schematically shows examples of modified balls by means of a special set G_L of functions which define the following equations

$$\begin{aligned} x_1 &= 0, & x_1 - 1 &= 0, \\ x_2 &= 0, & x_2 - 1 &= 0, \\ x_3 &= 0 \quad \text{and} & x_3 - 1 &= 0 \end{aligned}$$

in the Cartesian coordinate system $\{x_1, x_2, x_3\}$. The equations represents the boundaries of a cube.

The non-modified ball, as well as the modification, fulfill for all $X \in T_*$, the Axioms:

(V1) Each superset of a ball $U(X)$ is a ball of X.

(V2) The intersection of a finite number of balls $V(X)$ and T_* is a ball of X.

(V3) Each ball $V(X)$ contains X.

(V4) If V_1 is a ball of X then there exists a ball $V_2(X)$ that V_1 is a ball of any points $x \in V_2$.

The validity of the Axioms (V1) through (V4) with respect to the modified balls shows that the system of balls $U_*(X)$ represents a topology in each supporting set T_*. The arising topologies are called *trace topologies* or *induced topologies.* They arise by cutting the point set of the topologies by a subset of these topologies, for example as mentioned above by the supporting set T_*.Thereby, open sets are mapped to open sets and closed ones to closed ones.

We distinguish such a topology of a special T_* from another one, relative to the dimension of T_*, by the following relative notions:

Let $M \subset T_*$. Then $X \in M$ is called a *relatively inner* point if there exists a ball $U_*(X) \subset M$ and a *relatively frontier* point if each ball $U_*(X)$ contains at least one point of M as well as $T_* \setminus M$. Further we define

$int_* M$	the interior
$fr_* M$	the frontier
$cl_* M = int_* M \cup fr_* int_* M$	the closure
$ex_* M = T_* \setminus cl_* M$	the exterior

relative to T_*.

Definition 2.3. *Let T_* be a supporting set and let $M \subset E^3$ $(M \neq \emptyset)$ be an arbitrary set such that*

$$M_* = cl_* int_*(M \cap T_*).$$

Then we call the set M_* the (relatively) **closed regularization** *of M relative to T_*. The interior of a closed regular set is called the (relatively)* **open regularization**.

Relatively open regularizations are manifolds. It is possible to describe geometric objects by means of open as well by closed regularizations.

We defined the notion of a supporting set in the algebraic sense. The associated notion in the topological sense can be given as a regularization of algebraically supporting sets. The topological supporting sets only consist of a finite number of manifolds of the same dimension or their closure with respect to the kind of regularization desired.

A *region* of T_* is an open and connected set in T_* relative to T_*. A *domain* of T_* is a closed and connected set in T_* with respect to T_*. The domain $M \subset T_*$ is called *locally weak* in the point $X \in M$ if there is a neighborhood $U_*(X) \subset T_*$ and it holds for the set $W = M \cap U_*(X)$

$int_* W$ is not connected and $cl_* W$ is connected,

depending on T_*. X is called a *weak point.* Figure 2.2 shows a set M in the supporting set T_i $(i \in L)$ which contains a weak point X. The illustrated ball of X confirms this property.

The notions defined here will help us to describe polyhedra and their three-dimensional representations.

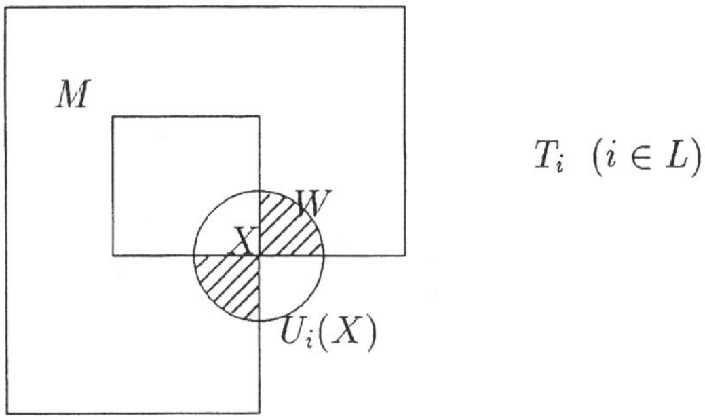

Figure 2.2. Example of a set M containing a weak point.

2.3. General and i-Polyhedra

Consider a surface $F_i \quad (i \in L)$ and the associated map $g_i \in G_L$ (a set of analytic functions). By means of the relation $g_i(x) = 0$, $g_i(x) > 0$ and $g_i(x) < 0 \quad (X \in E^3)$, a decomposition of the supporting set T_* into $H_*^=$, H_*^+ and H_*^- is given. The set $H_*^= \cap H_*^+$ is the *half space* of T_* according to g_i. It can be represented by the inequality $g_i(x) \geq 0$.

Definition 2.4. *A* **general polyhedron** $P = P(G_L, f)$ *is the result of a set operation by means of union, intersection and differences over a finite number of half spaces of* $T = E^3$ *according to a given operation formula* f.

Such polyhedra have some technically undesirable properties. Consider the set operations of polyhedra and choose all operands from the same dimension type; the result of the set operations will not necessarily be of the same type. Figure 2.3 shows set operations of the type $2D$ and $3D$, yet the result contains parts of lower types and they cannot clearly be classified by one and the same type.

Describing such properties of regularity more precisely, we have to introduce further notions of polyhedra. The dimension of a polyhedron is equal to the maximal dimension which arises from the combination of the half spaces. By regularization relative to the supporting set T_*, the dimension of the polyhedron does not change. Lower dimensional manifolds can only be added or omitted. Regular polyhedra are essential in representing general polyhedra. They even play an important role on the field of modeling.

Definition 2.5. *Each regularization* P_* *of a (general) polyhedron* P *relative to a supporting set* T_* *is called a* **simple polyhedron**. *A union of a finite number of simple polyhedra relative to one supporting set as well as to*

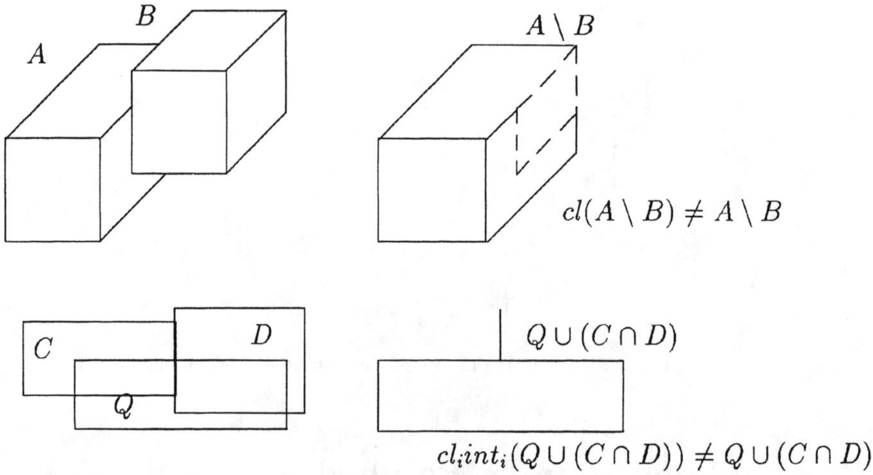

Figure 2.3. Set operations of regular sets of the same type need not produce a regular result.

different supporting sets of uniform dimension i is called an i-**polyhedron**. *The uniform dimension defines the dimension of* P_*.

A simple polyhedron is a i-polyhedron; i-polyhedra are regular sets. If we do not pay attention to the dimension and P is open or closed regularized, a simple i-polyhedron can be characterized by the equation

$$P = P_*$$

and P_* means the open or closed regularization of P, and its dimension can be equal to $0, 1, 2, 3$.

Definition 2.6. *A set of polyhedra and an associated connection formula of the desired set algebra, whose result is a general polyhedron* P, *is called a* **decomposition of the polyhedron** P.

Now we want to describe the topological properties of general polyhedra by means of i-polyhedra and the notion of the decomposition as mentioned above.

2.4. Special Algebraic and Topological Descriptions

Now we want to describe general polyhedra by parts of different dimensions and we want to connect those polyhedra with i-regularized operations. Firstly we want to connect the polyhedra in each of the dimension parts, and secondly we want to produce a common result. Therefore it will be useful to describe the general polyhedra by means of symmetric differences. The symmetric difference has a geometrically clear picture, see Figure 2.4.

We do not use those differences over objects of one and the same dimension however, we do use those differences for describing general polyhedra over parts of various dimension. Later we will see that it is easy to compute set operations algebraically.

If we regularize an arbitrary polyhedron P to P_* and assemble the added parts as well as the omitted parts of one dimension i into one and the same i-polyhedron in each dimension i, then the general polyhedron P can be represented as the result $p(0) = P$ of the recursion with four given regular polyhedra $P_i \quad (i = 0, 1, 2, 3)$

$$\begin{aligned} p(3) &= P_3 \\ p(i) &= p(i+1) \triangle P_i. \end{aligned}$$

In words, a general polyhedron $p(i)$ of dimension i is the symmetric difference ($\triangle$) between its regularization result P_i and a general $p(i+1)$ polyhedron of the higher dimension $(i+1)$. Such a recursion scheme is fixed by the four regular polyhedra P_0, P_1, P_2 and P_3. The recursion can easily be solved and it yields the following formula.

Definition 2.7. *A representation of a general polyhedron P*

$$P = P_3 \triangle P_2 \triangle P_1 \triangle P_0$$

by means of four regular polyhedra with different dimensions P_0, P_1, P_2 and

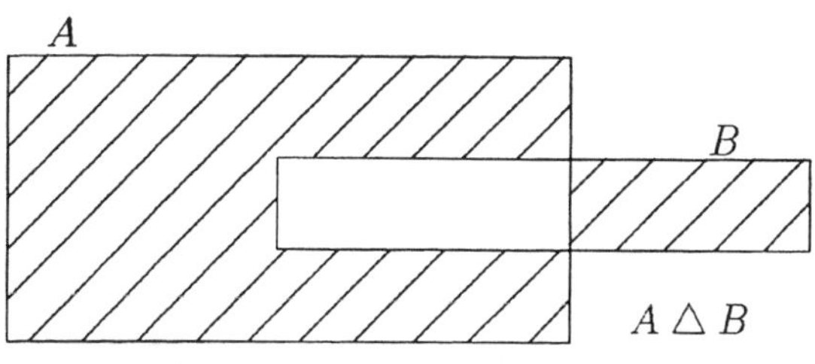

Figure 2.4. The symmetric diffenence of two rectangles.

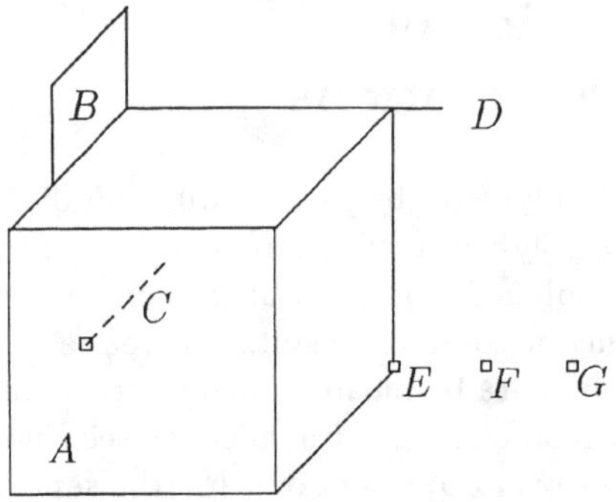

$$P_3 = A$$
$$P_2 = B$$
$$P_1 = C \cup D$$
$$P_0 = E \cup F \cup G$$

$$P = A\Delta(B\Delta((C \cup D)\Delta\{E, F, G\}))$$

Figure 2.5. Representation of a general three-dimensional polyhedron by symmetric differences of regular polyhedra.

P_3 and the symmetric difference (Δ) is called a **dimensions representation**.

It is possible that some of these four parts may be empty. Figure 2.5 shows a corresponding polyhedron that has a cube as the regular part P_* regarding to $T_* = E^3$.

Such a representation of a general polyhedron has special notions in the topological sense. Each of its representation parts consists of a finite number of pointwise connected sets of relatively inner points, the so-called (relatively open or closed) *connection components*. Furthermore, it is interesting to know which part of each describing polyhedron belongs to which polyhedra or its complement. Such maximally chosen parts are called *decomposition components.* The following equations define the decomposition of the four polyhedra.

$$\begin{aligned}
P_3 &= P_3 \\
P_2 &= (P_3 \cap P_2)^+ \cap (P_3' \cap P_2)^- \\
P_1 &= (P_3 \cap P_2 \cap P_1)^+ \cup (P_3 \cap P_2' \cap P_1)^- \cup (P_3' \cap P_2' \cap P_1)^+ \cup (P_3' \cap P_2 \cap P_1)^- \\
P_0 &= (P_3 \cap P_2 \cap P_1 \cap P_0)^+ \cup (P_3 \cap P_2 \cap P_1' \cap P_0)^- \cup \\
&\quad (P_3 \cap P_2' \cap P_1' \cap P_0)^+ \cup (P_3 \cap P_2' \cap P_1 \cap P_0)^- \cup \\
&\quad (P_3' \cap P_2' \cap P_1 \cap P_0)^+ \cup (P_3' \cap P_2' \cap P_1' \cap P_0)^- \cup \\
&\quad (P_3' \cap P_2 \cap P_1' \cap P_0)^+ \cup (P_3' \cap P_2 \cap P_1 \cap P_0)^-.
\end{aligned}$$

We denote the j-th term of P_i parenthetically by p_i^j. The signs $+$ and $-$ characterize an even and odd occurrence of negations ($'$) in the parentheses.

It is during computation of the representation formula, that p_i^j represents an added and a subtracted part if j is even or odd, respectively.

Now a special decomposition of an i-polyhedron P_i $(i = 0, 1, 2)$ is given by

$$P_i = \bigcup_{j=1}^{2^{(3-i)}} p_i^j \quad \textit{and} \quad p_i^j \cap p_i^k = \emptyset.$$

Fifteen partial polyhedra p_i^j for $i = 0, \ldots, 3$ and $j = 1, \ldots, 2^{3-i}$ arise which are pairwise disjunct or one of them contains the other. In a less restricted sense, we call any decomposition a *topological decomposition* if either the relatively inner points of each of the two decomposition parts are disjunct, or one of them contains the other part.

It is a problem that in the given decomposition of P_i the single parts p_i^j are not regular ones in general. If we regularize one of the decomposition parts p_i^j to P_i^j, then parts of lower dimensions, according to the representation formula, are influenced by the boundary of p_i^j. Those variations may be corrected in the lower dimensioned decomposition parts, and then it is possible to define topological decompositions of a regular kind. Moreover, the general polyhedron P can be represented by a difference of the sum over its positive decomposition parts and its negative parts. Therefore we distinguish between the following three representations.

Definition 2.8. *The decomposition of the general polyhedron P*

$$P = \bigcup_{i=0}^{3} \bigcup_{j \text{ even}} P_i^j \quad \setminus \quad \bigcup_{i=0}^{3} \bigcup_{j \text{ odd}} P_i^j$$

is called an **open decomposition** *or a* **closed decomposition**, *respectively, if and only if all of the (corrected) decomposition parts P_i^j are open ones and closed ones according to the dimension of P. It is called a* **mixed decomposition** *if and only if all of the even parts are closed ones and all of the odd parts are open ones according to its dimension.*

Figure 2.5 can be interpreted as a collection of relatively open as well as closed components, although it is easier to visualize the figure as a representation of relatively closed components. Indeed each representation of a general polyhedron can be given by means of relatively open as well as closed sets. A collection of relatively open sets is not equal to the same collection of closed sets. But each open representation of a general polyhedron can be transformed into a closed one and vice versa.

The dimensions representation and the topological decomposition of P explain their relations to each other from their very definitions. Each dimensions representation can be transformed into a topological decomposition of a special kind and vice versa. Another description often used for general polyhedra is the following one.

Definition 2.9. *A polyhedron representation by means of union and difference of a finite number of polyhedra is called an* **assembly**. *An assembly is called a* **topologically strong assembly** *if and only if each dimension component is decomposed topologically.*

Thus, if we consider the decomposition formula of a general polyhedron P from left to right, only those assembly parts are added, and so relatively inner points do not still belong to a partial result. Only those assembly parts are omitted whose relatively inner points already belong to a partial result. The decomposition of a general polyhedron can be assumed to be a topologically strong assembly. In this special case it follows

$$\bigcup_{i=0}^{3} \bigcup_{j \ odd} P_i^j \quad \subset \quad \bigcup_{i=0}^{3} \bigcup_{j \ even} P_i^j$$

and it holds for all possible i, k and even $j \neq l$ as well as only odd $j \neq l$

$$intP_i^j \cap intP_k^l = \emptyset.$$

In contrast to topological descriptions, we try to improve the idea of assembly representation of general polyhedra from the viewpoint of an algebraic application which will be covered later. As in the decomposition of a general polyhedron, it is possible to define a topologically strong assembly by dropping the assertion that all of the single regularization parts have to be either relatively closed or relatively open.

Definition 2.10. *A topologically strong assembly whose positive parts are relatively closed sets and whose negative parts are relatively open is called a* **mixed assembly**.

If we interchange the symmetric difference by unions and differences in the recursion scheme mentioned above and in the associated dimensions representation, then we get the recursion

$$\begin{aligned} p(3) &= P_3 \\ p(i) &= (p(i+1) \setminus P_i^-) \cup P_i^+ \end{aligned}$$

with

$$\begin{aligned} P_i &= P_i^- \cup P_i^+ \\ \emptyset &= int_i P_i^- \cap int_i P_i^+. \end{aligned}$$

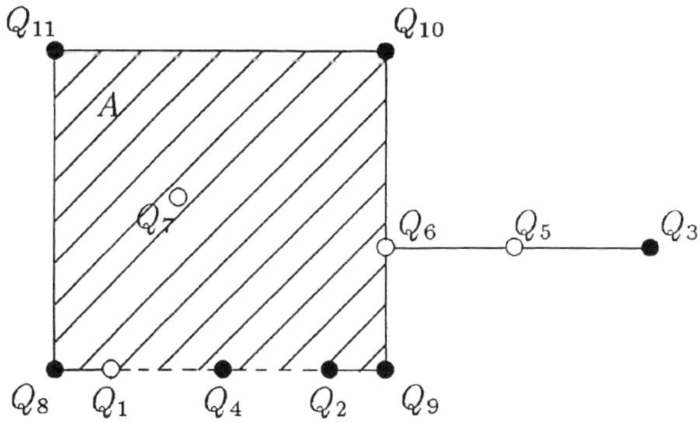

Figure 2.6. A Simple example that clarifies the differences of pure and mixed assembly representations.

If we can say correctly whether P_i^- and P_i^+ are to be closed or open, referring to the i-set, then the seven regular polyhedra $P_0^-, P_1^-, P_2^-, P_0^+, P_1^+, P_2^+$ and P_3 defining the general polyhedron P are well defined. The polyhedron P_3 is to be partitioned analogously into P_3^- and P_3^+.

Definition 2.11. *The eight i-regular polyhedra* $P_i^-, P_i^+ \quad (i = 0, 1, 2, 3)$ *represent the general polyhedron P by*

$$P = (((((((I \setminus P_3^-) \cup P_3^+) \setminus P_2^-) \cup P_2^+) \setminus P_1^-) \cup P_1^+) \setminus P_0^-) \cup P_0^+,$$

the so called **assembly representation**. *Here, I denotes either the fundamental set or the empty set.*

It is interesting to compare the different assembly descriptions with a simple example that shows only the differences, see Figure 2.6. The polyhedron P shown in this figure can be described by different closed and open dimensions representations of the form

$$P = P_2 \triangle P_1 \triangle P_0$$

with the pure closed or open regular polyhedra P_0, P_1 and P_2 and by different closed, open and mixed assembly representations of the form

$$P = (((P_2^+ \setminus P_1^-) \cup P_1^+) \setminus P_0^-) \cup P_0^+$$

by means of the corresponding positive and negative partial polyhedra of each dimension i. The sets I, P_3 and P_2^- are the empty set $\emptyset$ in the whole assembly representation. It yields the following descriptions. Thereby, the superscripts c and o denote the relative interior and the relative closure, respectively. The parameter of the function l denotes the vertex numbers of a polyline:

- Parts of the closed description

$$\begin{aligned}P_2 &= P_2^+ = A^c\\ P_1 &= P_1^- \cup P_1^+ = l^c(1,2) \cup l^c(6,3)\\ P_0 &= P_0^- \cup P_0^+ = \{Q_5, Q_7\}^c \cup \{Q_4, Q_2\}^c\end{aligned}$$

- Parts of the open description

$$\begin{aligned}P_2 &= P_2^+ = A^o\\ P_1 &= P_1^- \cup P_1^+ = \emptyset \cup (l^o(1,8,11,10,9,2) \cup l^o(6,3))\\ P_0 &= P_0^- \cup P_0^+ = \{Q_5, Q_6, Q_7\}^o \cup \{Q_4, Q_2, Q_3\}^o\end{aligned}$$

- Parts of the mixed description

$$\begin{aligned}\tilde{P}_2 &= P_2^+ = A^c\\ \tilde{P}_1 &= P_1^- \cup P_1^+ = l^o(1,2) \cup l^c(6,3)\\ \tilde{P}_0 &= P_0^- \cup P_0^+ = \{Q_1, Q_5, Q_6, Q_7\}^c \cup \{Q_4\}^c.\end{aligned}$$

It is worth noting that the mixed assembly representation has an essential property with respect to the regularization. If we regularize any mixed assembly representation in any dimension i, then the result is always a regular closed one in each dimension component k.

2.5. Surfaces, Curves and Vertices of Simple Polyhedra

Each (general) polyhedron can be represented as a connection of simple polyhedra. Therefore, we are especially interested in descriptions of simple polyhedra. The properties of the relative boundary of a simple polyhedron plays a fundamental role.

Let us consider: A simple polyhedron $P = P_*$ is given by a connection of half spaces defined by means of a set G_L of functions. The equations assigned to G_L divide E^3 into disjunct manifolds of different dimensions (solids, surfaces, curves and points in the topological sense). Each of these manifolds either belong to the relative boundary of the regular polyhedron $P = \bar{P}$ or they do not. And there is a connection of those manifolds belonging to the frontier of $P = P_*$ that together cover the whole boundary.

Definition 2.12. *The relative closure of the surfaces, curves and points of the frontier is called a* **boundary surface**, *a* **boundary curve** *and a* **boundary vertex** *of a simple polyhedron* $P = P_*$.

These boundary parts are subsets of algebraic surfaces, curves and points. They are again simple polyhedra and they are generally called *boundary polyhedra*. Recursively, boundary polyhedra of boundary polyhedra are called again boundary polyhedra of a simple polyhedron $P = \bar{P}$. The notions of a boundary curve and a boundary vertex are used in the same sense. Indeed, the notion of boundary areas need not necessarily be defined recursively in three-dimensional space. The following assertions show the foundation to describe simple polyhedra by means of its boundary.

Lemma 2.13. *The boundary of a simple i-polyhedron* $P_* = P(G_L, f)$ $(1 \leq i \leq 3)$ *is either empty or it is a* $(i-1)$*-polyhedron.*

Proof. The simple i-polyhedron P_* is a regular i-set. Each regular set has either no boundary (for example the surface of a ball) or a point Y exists and its whole ball $U_*(Y)$ lies in the associated supporting set outside of P_*. Furthermore, each regular set is either empty or it contains an interior point X and its whole ball $U_*(X)$. According to the construction of P_* and because of $(U1)$, the supporting set is divided by a non vanishing boundary of P_* into a finite number of regions that are homeomorphic to balls. These balls are i-dimensional and their frontiers are $(i-1)$-dimensional. It follows that the boundary of P_* is either empty or $(i-1)$-dimensional. This $(i-1)$-dimensional boundary is covered according to $(U1)$ by a finite number of (regular) parts with different supporting sets, i.e. by an i-polyhedron. □

Describing simple polyhedra, a different edge and vertex notion is often used. We shall make use of it as well.

Definition 2.14. *An intersection curve of two different boundary areas of a simple polyhedron* $P = P_*$ *is called a* **polyhedron curve**. *An intersection point of different polyhedron curves is analogously called a* **polyhedron vertex**.

The two edge and vertex notions differ from one another when it comes to weak polyhedra. A polyhedron is called weak if it contains at least one weak point. The two notations are equivalent if the polyhedron is not a weak one. Figure 2.7 shows three weak and simple polyhedra. The point U is a polyhedron vertex but not a boundary vertex. The weak polyhedron curve y in the second polyhedron is not a boundary curve. The curve z in

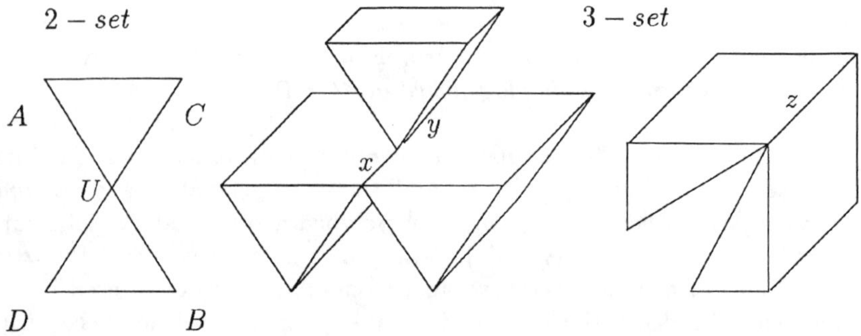

Figure 2.7. Examples of weak and simple polyhedra which shows different edges and vertex notions.

the third polyhedron is a weak boundary curve as well as x in the second polyhedron.

Furthermore it is to be stated that each point of a boundary curve belongs to at least one polyhedron curve because each boundary curve is at the same time an intersection of two boundary areas. The inverse of this assertion is not valid without loss of generality. Equivalent assertions can be stated regarding boundary and polyhedron vertices.

Indeed, weak polyhedra are sometimes interesting for representing technical objects, but they have to be considered in the implementations of set operations. Figure 2.8 shows two simple polyhedra which are to be

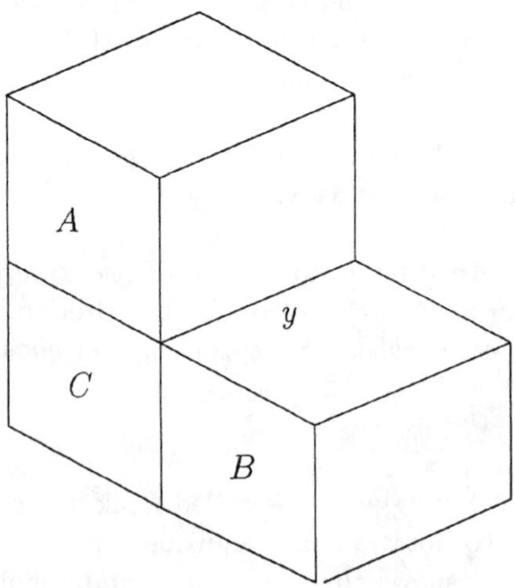

Figure 2.8. The partial result $A \cup B$ of the regular union $A \cup B \cup C$ has a weak polyhedron line y, although A, B, C and the final result are not weak ones.

connected. Depending on the computation order of the union $A \cup B \cup C$, the line y is weak in the partial union $A \cup B$ but does not even exist in the other partial result $B \cup C$.

The boundary of a simple polyhedron $P = P_*$ divides the supporting set T_* of the polyhedron into three sets of interior, exterior and boundary points. The *orientation* of the boundary of $P = P_*$ defines which of the two other sets (disjunct to T_*) is the set of inner points of $P = P_*$. It can be given by the location of a special, relatively inner point or by defining the orientation (gradient) of the supporting set of each boundary piece in the same sense.

2.6. Boolean Characterization of Oriented Boundaries

The regularized parts of a general polyhedron can be defined either by means of CSG-representation by systems of inequalities and a connection formula, or by a boundary representation. To transfer an inequality representation into a boundary representation as well as to connect boundary representations in a Boolean kind, a special local homeomorphic map is now presented. It characterizes the location of any point $X \in E^3$ regarding a Boolean connection of oriented boundaries.

Let us consider the supporting set T_* of a regular (closed) polyhedron P of dimension $d \geq 1$. The symbol T_l denotes a supporting area that is linearly independent on the supporting set T_*. H^+ and H^- are the positive and negative parts of the associated half space of T_l. Γ_{*l} defines the union of all supporting sets with the exception of $T_* \cap T_l$.

Definition 2.15. *A function λ of a point $X \in \mathcal{L} = T_{*l} \setminus \Gamma_{*l}$ with $\lambda(X) \in \{T, F\} \times \{T, F\}$ and*

$$\lambda(X) = \begin{cases} (F, F) & X \in cl_* P \\ (T, F) & X \in rd_* P \text{ and } U \cap H^+ \cap P \neq \emptyset \\ (F, T) & X \in rd_* P \text{ and } U \cap H^- \cap P \neq \emptyset \\ (T, T) & X \in int_* P \end{cases}$$

is called a **location characterization function** *dependent on the regular polyhedron P, the supporting set T_* and the supporting area T_l and by means of a sufficiently small ε-ball U of the considered point X.*

The letter F means the logical value $.false.$ and the letter T means $.true.$. The set $\mathcal{L}$ consists of regions with only a constant location function. The

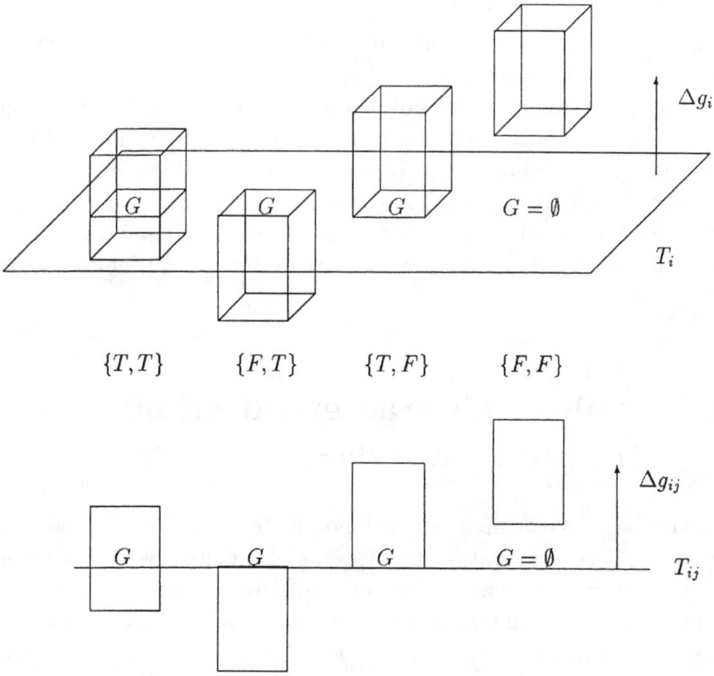

Figure 2.9. Classification of supporting regions that are intersections with polyhedra.

relative closure of $\mathcal{L}$ is the supporting set $T_* \cap T_l$. The following boundary criterion can be used in algorithms.

Lemma 2.16. *A region $G \subset T_* \cap T_l$ is a part of the boundary polyhedron P_l of the regular polyhedron P in T_l if and only if the location function λ has the value (F, T) or (T, F) in one point $X \in G$.*

Each point X of a supporting set can be classified according to the location of half spaces or whole (regular) polyhedra. Figure 2.9 shows the classification of points from G of the intersection of a cube with a supporting set T_i and T_{ij} respectively, and the associated classifications.

We are now interested in a relation of region classification between a set of polyhedra that belong to a formula description and the associated set operation result. By means of the following lemma it is possible to compute the location function of a Boolean set operation with the location function of each polyhedron operand.

Lemma 2.17. *Let $A_1, \cdots, A_n$ be regular polyhedra and let $f(A_1, \cdots, A_n)$ be an associated Boolean connection with the location function $\lambda(X)$. For the single location functions $\lambda_i(X)$ of the operands $A_1, \cdots, A_n$ $(i = 1, \cdots, n)$*

$$\lambda(X) = \tilde{f}(\lambda_1(X), \cdots, \lambda_n(X))$$

holds where $\tilde{f}$ represents the same Boolean term as f in the first component as well as the second one of λ_i independently.

A homeomorphic map is given by this relation of λ between the describing set algebra and the classification algebra. After computing the location function $\lambda_1(X), \cdots, \lambda_n(X)$ of all describing polyhedra in the point X, the common classification $\lambda(X)$ can easily be built. We get the two different edge notions directly from their definitions as follows.

Lemma 2.18. *A region G that belongs to a supporting area with location function λ^1 and to a supporting edge with location function λ^{12} is a part of a boundary edge if and only if for any $X \in G$*

$$\lambda^1(X) = \{T, F\} \vee \{F, T\} \quad \textit{and} \quad \lambda^{12}(X) = \{T, F\} \vee \{F, T\}.$$

A region G that belongs to a supporting area with the location function λ^1 and to another linearly independent supporting area with the location function λ^2 is a part of a polyhedron edge if and only if for any $X \in G$

$$\lambda^1(X) = \{T, F\} \vee \{F, T\} \quad \textit{and} \quad \lambda^2(X) = \{T, F\} \vee \{F, T\}.$$

An analogous assertion about boundary vertices and polyhedron vertices can be concluded.

Because of the finite number of possible regions of one supporting set, it is only necessary to classify a finite number of points. If we define polynomial equations by G_L with a higher degree, it is still necessary to test the dimension of algebraic variety of the investigated regions. A decrease in dimension destroys the assumption of the criteria. The test of the dimension can be done by means of the Jacobian matrix.

Most modelers with Boolean set operations can only connect the operands pairwise, i.e., a multiple set operation has to be reduced to binary operations. This is always possible.

Another argument is time complexity. Although Boolean set operations are problems with polynomial time complexity, they need some more combinatoric ideas. If we describe geometric objects not by pairwise connections but by a multiple operation, then it is necessary to find suitable tests which can easily separate disjunct polyhedra. Only in that way, the running time of set operations seems to be small enough.

Transformation algorithms from CSG- into boundary representation need the characterization of their polyhedron edges and vertices. The multiple operation is better than the pairwise connection in those algorithms because the programs are shorter.

3

Boundary Models

An i-polyhedron can be represented by simple polyhedra. The simple i-polyhedra can be represented by their oriented supporting set of dimension i and their $(i-1)$-polyhedra of the boundary, etc. The boundary polyhedron characterizes the i-polyhedron because its orientation indicates which side of the supporting set faces interior and exterior. Such polyhedra curves and vertices, which are not boundary curves and vertices could additionally be used. Taken into the data model too, then there is no reason to distinguish them from the boundary curves and boundary vertices separately.

3.1. Tree and Graph Structures

The *tree structure* naturally describes the i-polyhedra. It contains a list of simple polyhedra which describe all i-polyhedra of an object. Each simple polyhedron is described by its supporting set and pointers to simple polyhedra of lower dimension which characterize the boundaries, and so on. If some pointers point to the same polyhedron, these pointers have to be doubled in a unique sense so that a tree of pointers arises.

This tree structure can be very beneficial. The advantage is based on its simplicity: The relative boundary parts are ordered from the root to the leaves according to their dimension. This tree structure is, in this sense, enough to store boundary representations of i-polyhedra.

Clear disadvantages of this simple structure exist too. For example, a vertex of a cube belongs to three boundary surfaces and six boundary curves. Therefore it arises three times and six times, respectively, in the tree structure mentioned above. But such repetitions are unfavorable in the implementation of combinatorial algorithms. If such an algorithm needs all incident surfaces to one given vertex during computation, the tree must be searched, causing an unjustifiable temporal expense.

If you want to use the internal representations of polyhedra for implementing algorithms with high combinatorial expense, then a so-called *complete structure* has to be used. In contrast to the tree structure, each polyhedron (surface, curve or vertex) is stored exactly once. The weak polyhedral curves and vertices are an exception. They can be stored more than once according to their incidence in order to distinguish surface pieces and curve pieces, which are separated by weak polyhedra.

The structure is completed by adding inverted pointers. The relationship between two neighboring nodes in the tree can be found without tree searching. Thus the set of all pointers gives the information of all incidence relations between surfaces, curves and vertices.

Element modifications in both structures mentioned above cannot be realized so easily because the change of an element is only possible by changing other pointers more than once. It is unfavorable to use algorithms which modify the data structure. In contrast to the data structures which explicitly address the geometric elements, it is also favorable to work by means of a so-called *implicit structure*. The fundamental idea behind implicit structures for describing boundary representations is based on using special graph structures. Here, the pointers are essentially eliminated. The references are implicitly stored in the order of the information. It allows compact storage and fast algorithms. In particular, these structures are most advantageous if the polyhedra are to be approximated by means of linear functions. Although, they are advantageous for curves and surfaces of higher degree too.

3.2. Special Structures of the Graph Level

Let us consider three graph structures which can be taken as a base for describing an implicit structure. They are classified by using special basic elements in the *surface structure*, *edge structure* and *vertex structure*, and they behave differently with respect to incidence and combinatorial aspects. At first we simplify these three structures using only linearly restricted polyhedra and later (Chapter 4.4) we discuss the usefulness of these structures by describing other, more general polyhedra.

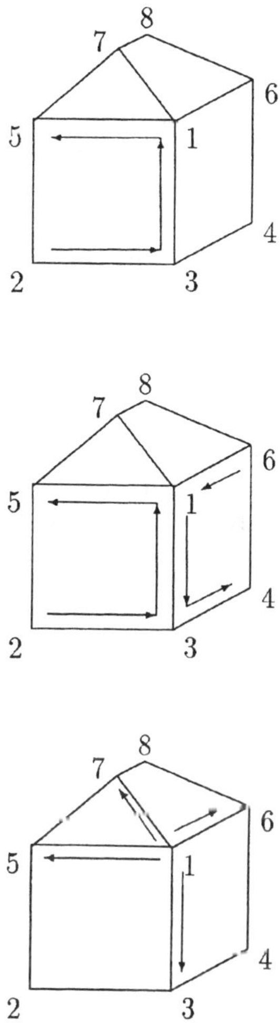

surface structure

$F1$: [1 5 2 3 1]
$F2$: [4 6 1 3 4]
...

edge structure

$E1$: [2 | 3 1 | 5][6 | 1 3 | 4]
$E2$: [5 | 1 7 | 5][8 | 7 1 | 6]
...

vertex structure

$V1$: [6 7 5 3 6]
$V2$: [3 5 9 3]
...

Figure 3.1. Surface, edge and vertex structures of a simple example.

Each of the three structures are based on a point list that consist of all vertices exactly once. The location index of the coordinates, corresponding to a vertex in the list, defines the vertex number. Furthermore, ordered index sets of points are enumerated with different significance. Figure 3.1 explains these three structures.

The marked surfaces show the implicitly defined incidence that is stored in each list element ($F1, F2, \cdots; E1, E2, \cdots; V1, V2, \cdots$) by the location of the vertex number in the ordered index sets. Depending on the application, each structure is important. The surface structure is advantageous for production engineering and the representation of geometric objects. The structure has a combinatorial disadvantage. It stores only implicitly the edge incidence within a boundary polygon of a surface. The incidences of the surfaces among one another are not known without expensive searching.

The edge structure partially eliminates this disadvantage. The incidence between two surfaces can be recognized easier. Set operations and hidden line algorithms of described objects are easy to implement. Such structures are well-known and were used in the first modeling algorithms. But they do not reflect all possible incidences among elements of the graph level without a special graph searching algorithm. For instance, the edge incidences (see the edge from 1 to 7 in Figure 3.1) are not known directly. An advantage, in contrast to the other two structures, is the constant length of the elementary information. The initial elementary information, in a common list, can easily be computed depending only the edge number. In the same way, cancellations of old edges and definitions of new edges are simple operations during any computation algorithm.

The vertex structure has very interesting combinatorial properties. It represents all possible implicit incidences of the graph. Indeed, it is necessary that each vertex number is equal to its location index in the edge list.

The surface and vertex structures have more extensive list organizations like the edge structure. If a modeler has a CSG-structure, the implemented CSG-organization is able to receive the list organization of the surface or vertex structure.

All three graph structures are useful to represent different dimensional i-polyhedra. Then, if it is necessary, edges and vertices have to be defined as a degeneration of surfaces and edges, respectively (see Figure 3.2).

In each of the three structures, the incidences are entered corresponding to the orientation of the polyhedron representation (right-hand-rule) with at most one exception. Without loss of generality, the edge structure does not have the possibility to admit more than two surface incident to one edge because of the fixed length of the list atoms. In this case, it is useful that an incidence of more than two surfaces is not recognized. Figure 3.3 shows an incidence of an edge to three surfaces by means of a 2-polyhedron. The surface structure and the vertex structure are nondegenerate. In the edge structure, the corresponding pointers at the edge of number eight are omitted.

The example also shows the degeneracy of the boundary edge $(2,4)$ of the surface $[2 \quad 3 \quad 4 \quad 6 \quad 2]$ by containing the point 6. It is advantageous to divide those edges like $(2,6), (6,4)$. The vertex structure then gives at least the complete local information of the entire incidence.

a 2-polyhedron of the E^3

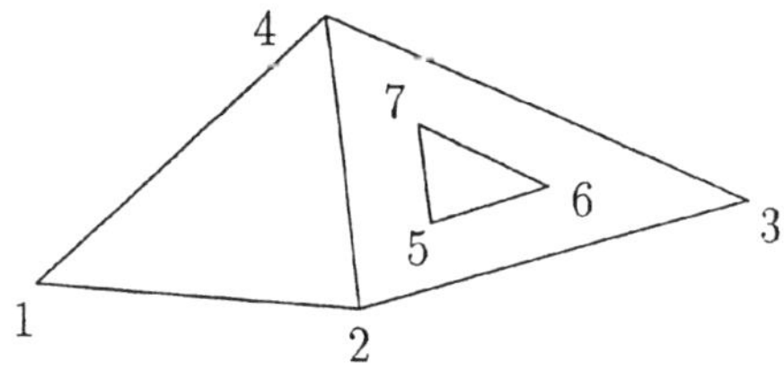

a 1-polyhedron of the E^3

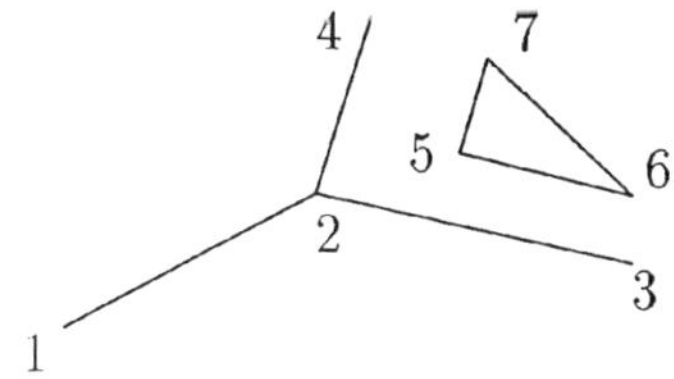

surface structure
$F1$: [1 2 4 1]
$F2$: [4 2 3 4][7 6 5 7]

edge structure
$E1$: [1 2 4 1][3 4 2 3]
$E2$: [4 1 2 4][* * * *]
$E3$: [4 2 3 4][* * * *]
$E4$: [2 3 4 2][* * * *]
$E5$: [5 6 7 5][* * * *]
$E6$: [7 6 5 7][* * * *]
$E7$: [6 5 7 6][* * * *]

vertex structure
$V1$: [2 4]
$V2$: [3 4 1]
$V3$: [4 2]
$V4$: [1 2 3]
$V5$: [7 6]
$V6$: [5 7]
$V7$: [6 5]

degenerate surface structure
$F1$: [1 2 4 *][2 3 *]
$F2$: [5 6 7 5]

degenerate edge structure
$E1$: [1 2 4 *][* * * *]
$E2$: [* 1 2 4][* * * *]
$E3$: [4 2 3 *][* * * *]
$E4$: [5 7 6 5][* * * *]
$E5$: [7 6 5 7][* * * *]
$E6$: [6 5 7 6][* * * *]

degenerate vertex structure
$V1$: [2]
$V2$: [1][3][4]
$V3$: [2]
$V4$: [2]
$V5$: [6][7]
$V6$: [5][7]
$V7$: [5][6]

Figure 3.2. Surface, edge and vertex structures describing i-polyhedra of different dimensions. The examples shown in the last two lists demonstrate the list degeneration describing lower dimensional i-polyhedra. The symbol * means that a successor or a predecessor does not exist. The edge structure is bad in this case because much of the storage is empty and, moreover, information has to be stored relatively often.

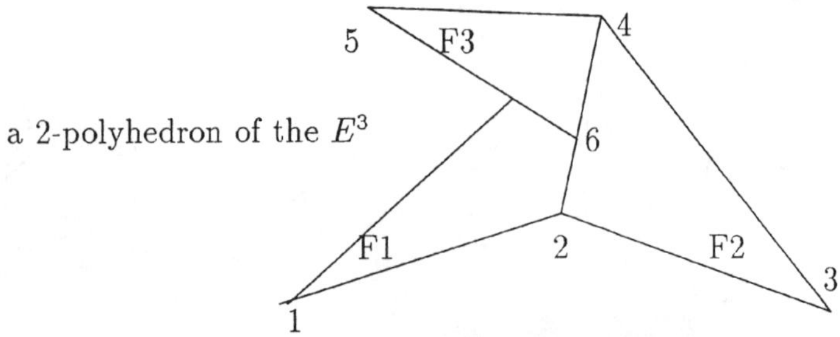

surface structure
$F1$: [1 2 6 4 1]
$F2$: [2 3 4 6 2]
$F3$: [4 5 6 4]

vertex structure
$V1$: [2 4]
$V2$: [3 6][6 1]
$V3$: [2 4]
$V4$: [6 3][5 6][1 6]
$V5$: [6 4]
$V6$: [4 5][2]

edge structure
$E1$: [4 1 2 6][* * * *]
$E2$: [1 2 6 4][4 6 2 3]
$E3$: [2 6 4 1][3 4 6 2]
$E4$: [6 4 1 2][* * * *]
$E5$: [6 4 5 6][* * * *]
$E6$: [4 5 6 4][* * * *]
$E7$: [5 6 4 5][* * * *]
$E8$: [6 2 3 4][* * * *]
$E9$: [6 2 3 4][* * * *]

Figure 3.3. Implicit description of a 2-polyhedron in E^3 containing the edges $(6, 4)$ with three incident surfaces.

3.3. Linear-Time Transformations of Graph Structures

To make the entire graph incidence available for composing algorithms, we now try to transfer the surface structure and the edge structure to the vertex structure. The principle in both algorithms is the same. The surface structure and the edge structure are passed three times sequentially.

Principle of the Algorithm

(Transformation of the surface or edge into the vertex structure)

Step 1: Count the edges which begin at each vertex and compute the pointer describing the beginning of each edge element in a common list.

Step 2: Each edge (i, j) is read and the edge j is stored at the i-th pointer. Then the i-th pointer is moved on.

Step 3: Sort of the list elements according to the surface incidence only within each single vertex list atom.

If n is the number of edges then the first two steps obviously need $O(n)$ operations (that means the expense grows proportionally to n). Step 3 is a little more difficult. According to the input of the starting list, the vertex $V1$ in Figure 3.1 can be described after the first two steps by pairs of neighbor vertices

$$(3,5),\quad (6,3),\quad (5,7),\quad (7,6)$$

without any ordering. How do we obtain the desired combination? Without loss of generality, we can assume that the number of neighboring vertices of a vertex differs from one another. Now we put this value into a one-dimensional array A at the place i with respect to all the given edges pairs (i, j). Simultaneously we create a vector B, where we have made notations in A.

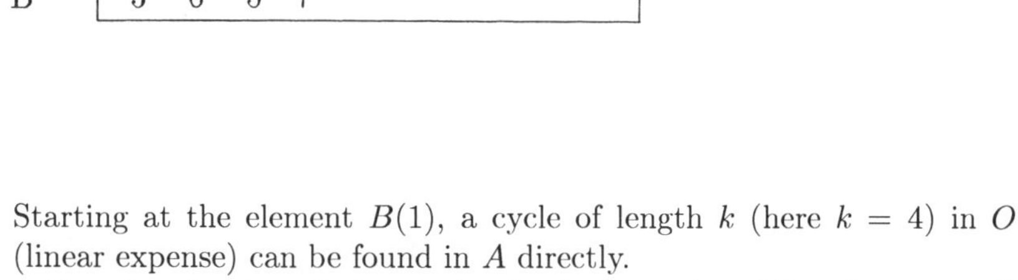

Starting at the element $B(1)$, a cycle of length k (here $k = 4$) in $O(k)$ (linear expense) can be found in A directly.

Crossing the described ordering in A, the assigned list elements will be overwritten by the marker *. If the loop stops at the end of cycle searching, a new start begins with the next new element of B. If $A(B(2))$ is already marked, then $B(3)$ is the next start and so on.

Finally, this algorithm permits the neighbor searching of an ordering according to the assigned incidence surface - edges - surface with $O(k)$ - Step 3 and:

Summary. *If n is the edge number of a surface or edge structure describing an i-polyhedron, this structure can be transformed into the vertex structure by the expense of $O(n)$.*

The inverse transformations are important, too. They are not always possible without searching. To produce the edge structure from the vertex structure, the predecessor of an edge (i, j) can be read directly from the vertex structure. The successor has to be searched for on the vertex j. Even so, the edge (j, i) has to be searched for from the vertex j and the predecessor and the successor must be defined analogously to (i, j).

Such a search can already effect $O(n)$ operations in the worst case, causing the total expense to be linear. It depends on the number of edges which start from a vertex. In practice, this number is very small on average and therefore it can be estimated by a suitable average constant. Seldom do many edges begin from one vertex. In this case, there are typically many vertices that have a small number of edges (see for example a linearly approximated cone). One has therefore the impression during computation that the common transformation takes a linear expense.

The above discussion holds in the transformation of the vertex structure to the surface structure: The successor must be found in each case of an edge (i, j) for the vertex i and the vertex j. The goal is to get linear time during computation. This method finds the necessary cycles describing a surface structure. Although it cannot be decided which of the cycles belong to one and the same surface.

It is necessary to ask for their associated equations, and which of two oriented cycles, of one and the same surface equation, is a relative inner cycle of the other. The arising expense cannot be clearly predicted because it depends obviously on the shape of the polyhedron.

In the special case that the surface and the boundary surface are planar and convex, each cycle is the boundary of a surface. An additional assignment of disjunctive cycles is not necessary anymore. Then each of the three structures described above can be transformed into one another in nearly linear time and the graph level is such that all incidences can be determined in a sufficiently small time.

One can also say that the topological as well as the combinatorial advantages are sufficient to satisfactorily describe geometric objects by planar and convex boundary surfaces. Such descriptions obtain, in practice, particular significance for representation schemes of the computer internal data structure, for approximations of non-linear surfaces, or for the generation of grids for the finite element method.

3.4. A Possible Common Structure of i-Polyhedra

The graph structures mentioned above permit their generalization for i-polyhedra without linear restricted boundaries. The following symbols help us to describe such a possible structure. This structure is complete and uses the advantages of implicit definitions (see Scheme 3.4):

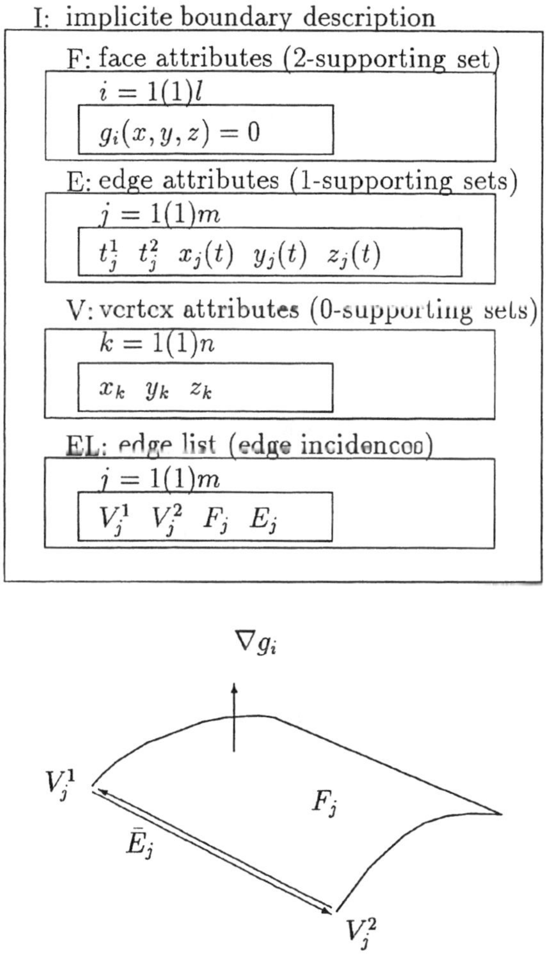

Scheme 3.4. Edge structure describing a boundary representation of a i-polyhedron.

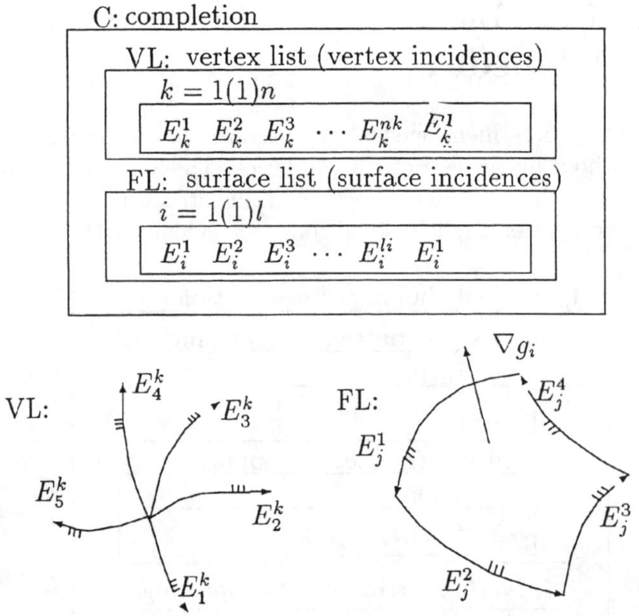

Scheme 3.5. Completion of the structure by vertex and surface structure.

i, j, k	number of the surface, edge and vertex
g_i, G_L	function of the analytical description of the t-th surfaces
∇g_i	its assigned gradient
t_j^1, t_j^2	number of starting and end point of the edge j
$x_j(t), y_j(t), z_j(t)$	parametric representation of the edge j
x_k, y_k, z_k	coordinates of the point k
V_j^1, V_j^2, F_j	starting and end point and the bounding surface of the edge j
$\bar{E}_j$	number of the edge contrary to edge j.

This structure is used to store objects because of the simplicity of its edge list. Often the desire arises to work during computation by means of the vertex and surface structure. The given edge structure permits the expense, see Scheme 3.5. It means

E_k^r the number of the r-th edge with starting point k $(r = 1, 2, \cdots, n_k)$

E_i^k the number of the k-th edge of the surface i.

Polyhedron algorithms produce extensive data. Modeling algorithms connecting polyhedra react very sensibly to input errors. However, the ability must exist to recognize and compensate for input errors. Implicit data structures permit fast so-called *confidence tests* to prove the incidence correctness. In this sense it is favorable to describe 3-polyhedra to double the weak polyhedron edges in the data structure such that there exists exactly two surfaces incident to each edge. Even the weak point in 2-polyhedra should also be doubled that exactly two edges exist to each vertex of the implicit description. Without searching, the correctness of the incidences can be examined.

3.5. Half-Space Model into Boundary Representation

Some systems of inequalities of points from E^3 and their connection by means of a set formula defines a so-called *half-space model* of a polyhedron, see Figure 3.6. Assuming that a special regularization type is given, the half-space model can be taken to describe simple polyhedra, i-polyhedra as well as general polyhedra. The regularization type tells us which of the points in E^3 have to be omitted or added to the solution points of the half-space description.

Sometimes it is necessary to transform the half-space model into a boundary representation. Therefore, the unknown edges of the boundary representation are needed. We intersect all of the restricting faces of the given half-spaces, and determine whether the intersection curve pieces are parts of the resulting edges. Without a combinatorial idea, it leads to an algorithm with a high complexity, i.e. a long run-time.

If we use linear systems of inequalities, then each system represents a convex polyhedron. A modification of the well-known simplex methods, which solves linearly restricted optimization problems, will show us a practically sufficient combinatorial idea.

The Simplex

Let us consider the matrix $A = (a_{ij})_{3,n}$ and the vector $b = (b_i)_n$. The set S of all elements $x \in R^3$ with

$$0 \leq A^T x_N + b \tag{3.1}$$

is called a *simplex*. For our purposes and without loss of generality, the boundary of set S is finitely restricted. The system (3.1) is transformed by

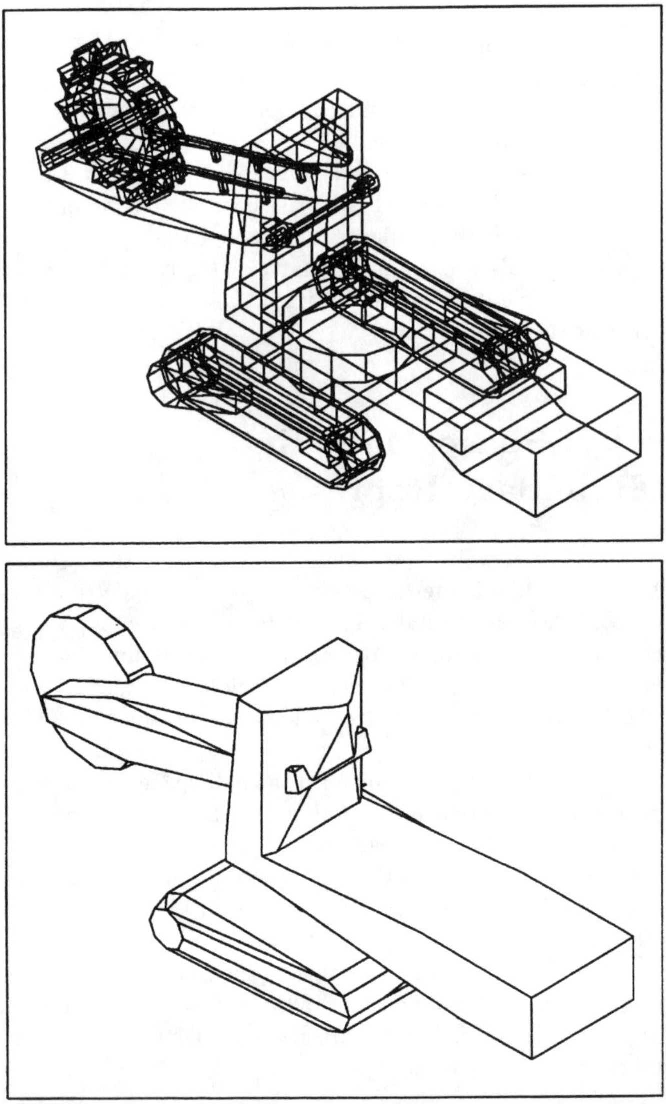

Figure 3.6. The boundary representation of a linearly restricted half-space model (CSG) and a geometric abstraction of the boundary model by means of convex hulls.

an additional vector $x_B \in R^n$, $x_B \geq 0$ of variables into the *basic description* of S

$$x_B = A^T x_N + b \, , \quad x = \begin{pmatrix} x_N \\ x_B \end{pmatrix} . \tag{3.2}$$

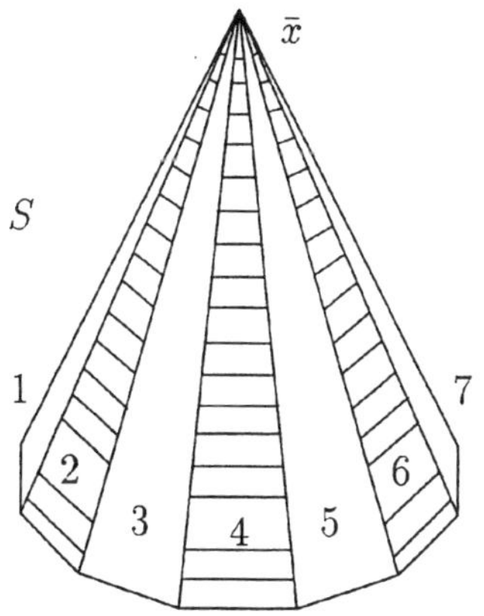

$$N = \{2, 4, 6\}$$

and

$$1, 3, 5, 7, \cdots \in B$$

Figure 3.7. A possible basic description of the cone top.

The vector x_N is the vector of the non-basic variables, x_B the vector of the basic variables, N is the set of the non-basic indices and B is the set of the basic indices. Such a simplex $S \neq \emptyset$ of the R^3 is a convex body, a convex area, a straight line or a point. In the following, we discuss two of the elementary questions about simplices in order to describe the boundary representation of the simplices themselves.

The vector $\bar{x} = (0 \quad b)^T$ is called the *basic solution.* Basic descriptions are equal to each other if their equations are respectively pairwise proportional. If we exchange the variable r $(r \in B)$ with the variable s $(s \in N)$ by means of the usual rules for transforming a system of equations into another equivalent system, then a new basic description arises. The element a_{rs} is called a *pivot* and the transformation is called an *exchange.* A basic solution $\bar{x}$ is called the *corner* if $\bar{x} \geq 0$. At the same time it also characterizes, in a corner of the simplex S, the point set itself.

A basis description is characterized by the set N of its non-basic indices. The following example shows the problem of the non-injective assignment of a basic solution to the basic description.

Example 3.1. A cone of the space R^3 is defined by the intersection of half-spaces. This is a simplex S, see Figure 3.7.

The hatching signifies the planes of the basic description (N) of the basic solution $\bar{x}$. The corner $\bar{x}$ belongs to the border planes 1 through 6, i.e. $x_i = 0$ for $i \in \{1, 2, 3, 4, 5, 6, 7, \cdots, 12\}$. Specifically, this is valid for $i \in N$. The basic description (N') with $N' = 2, 3, 4$ is also a basic description of $\bar{x}$.

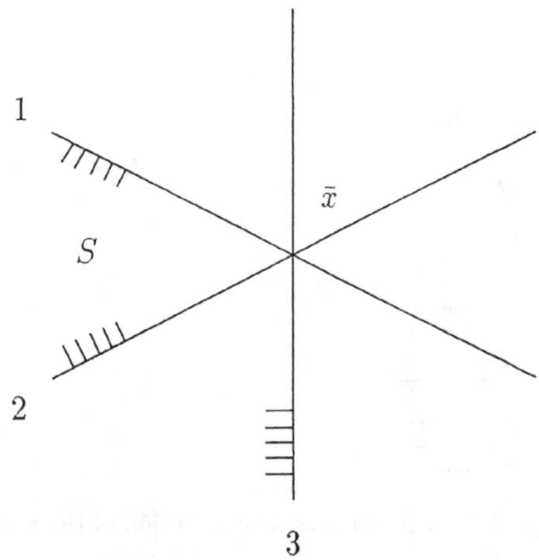

Figure 3.8. The inequality 3 produces a $\bar{x}$-active but redundant half-space for the plane simplex S.

The essential question is whether one can reduce a basic description (3.2). If (3.2) yields the system of inequalities (3.1) by omitting the inequality $x_i \geq 0 \;\; i \in N \cup B$ with the new solution set S' and

$$S = S'$$

then the inequality i is called a *redundant inequality.* A redundant inequality with $i \in B$ is called a *redundant of the basic description* (3.1) if from $x_N = 0$ follows $x_i = b_i = 0$.

A redundant inequality can be omitted if it is detected. The basic description (3.2) reduces in this way. We do not change the notation. Nevertheless no redundant inequalities exist in Example 3.1, as the basis description can multiply define the corner $\bar{x}$ by the intersection of planes.

Furthermore, a basic description can be reduced if the simplex S belongs to a plane and a line, respectively. Then the attached equation, which defines the plane or the line, can be substituted.

If S is planar, then by deleting redundants, it is possible to assign the basic description and the basic solution injectively. Let us consider the plane containing S and three half-spaces whose boundaries are incident to the basic solution $\bar{x}$. Because we consider the intersection of half-spaces, all inequalities of the type 3 in Figure 3.8 are clearly redundant ones.

After their cancellation, only exactly two of these inequalities remain because $S \subset S'$ and $int_i S = \emptyset$. Then the index set $N = \{1, 2\}$ defines the basic solution $\bar{x}$ as the intersection of the half-space boundaries well.

The enumeration of the simplex corners is important. Thereby we will use the basic descriptions to produce boundary representations. Because of the reasons mentioned above in Example 3.1 and looking at more complex examples, we will concentrate on the planar simplex in R^3. If we reduce the attached basic description then we can take the interpretation as a simplex in R^2.

Enumeration of the Simplex Corner in R^2

It is important for complex geometric examples in R^3 to enumerate the corner of a boundary simplex $S \subset R^2$. With this aim, we add a linear weight function to the basic description (3.2)

$$\min \quad c^T x_N + c_0 \tag{3.3}$$

$$x_B = A^T x_N + b \,, \quad x = \begin{pmatrix} x_N \\ x_B \end{pmatrix} \tag{3.4}$$

with any vector $c = (c_i)_3 \neq 0$ and c_0.

Let us look now for the corners of a planar simplex:

Solving linear minimization problems, the simplex method is used. We reduce the optimization problem (3.3) of R^3 by means of a supporting plane S, to a problem in R^2 and consider some steps to enumerate the corners.

Computation of a new corner:

Because of the characterization theorem of the linear optimization, the basic solution $\bar{x}$ is a suitable one (i.e. it is a corner) if $b \geq 0$. We choose $c_u = 1$ and c_s with $u, s \in N$, $s \neq u$ and $c_0 = 0$. A change of the pivot element p_{rs}, that was chosen by the pivot rule,

$$q(s) = \min_{i \in B} \{-\frac{b_i}{a_{is}} \mid a_{is} > 0\} \tag{3.5}$$

leads to a basic description whose basic solution is also a corner. If the old basic description does not have a redundant equation then no $s \in B$ exists in R^2 that $q(s) = 0 \quad (b_i \neq 0$ for all $i \in B)$. That is the function value yields, after the change

$$\bar{c}_0 = c_0 - \frac{b_r}{a_{rs}} c_s = c_0 + q(s) c_s \neq c_0,$$

and in this way we not only get a new basic solution but a new corner, too. In the following step, the new direction change will be fixed by defining of $c_u = -1$ and $c_s = 1$ and so on.

The choice of the best new basic description:

If the characteristic quotient $q(s)$ is not obvious, then we use a second one. Therefore, we define the vector $x_N = x_n + \varepsilon$ by means of a suitable ε-vector ($\varepsilon \geq 0$). The basic description of (3.3) with respect to $\bar{x}$ is transformed into

$$x_B = A^T x_N + (b - A^T \varepsilon) .$$

In particular, we choose $\varepsilon_s = 0$ and $\varepsilon_u \geq 0$ for $s \neq u$ and $s, u \in N$, see Figure 3.9.

Let us consider the index set Q

$$Q = \{i \in B \mid -\frac{b_i}{a_{is}} = a(s)\}$$

as the set of the indices for which the attached pivot is equal to the characteristic quotient $q(s)$. Also let us compute the characteristic quotient according to (3.5) with the problem changed by the ε-variation over the domain $B \cap Q$. We get

$$t_1(s) = \min_{i \in B \cap Q} \{-\frac{b_i - a_{iu}\varepsilon}{a_{is}} \mid a_{is} > 0\}$$

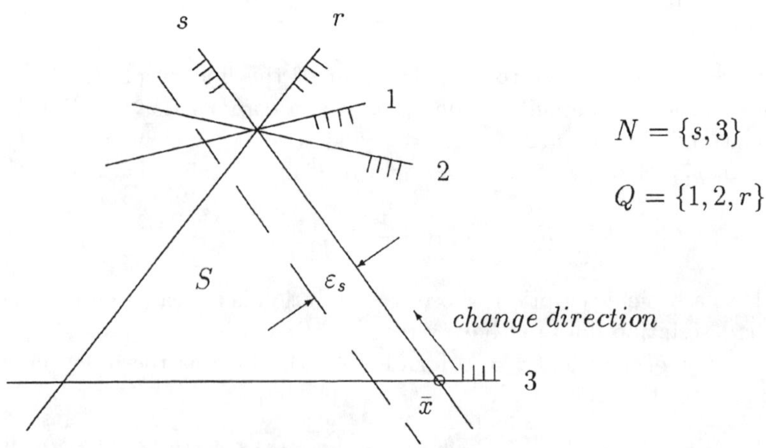

Figure 3.9. The ε-problem to get a well-defined basic description.

and let us define the following second characteristic quotient

$$\bar{q}(s) = \frac{1}{\varepsilon}(t_1(s) - q(s))$$

or

$$\bar{q}(s) = \min_{i \in B \cap Q} \{\frac{a_{iu}}{a_{is}} \mid u \in N, u \neq s\} \ . \tag{3.6}$$

This quotient chooses, from the set Q, exactly one index r: if a $r' \neq r$ exists with

$$\bar{q}(s) = \frac{a_{ru}}{a_{rs}} = \frac{a_{r'u}}{a_{r's}} \quad and \quad q(s) = -\frac{b_r}{a_{rs}} = -\frac{b_{r'}}{a_{r's}}$$

then the equality of the associated half-space restrictions follows directly. Also it follows that the new basic description of the ε-problem is well-defined. The quotient $\bar{q}(s)$ is independent of ε. With $\varepsilon = 0$, the desired basic description is well-defined too. All of the half-space restrictions described by the index set Q, and apart from the special pivot index $r \in Q$, are redundant and can be omitted.

The well-defined dual step:

In starting the enumeration of the corners by means of pivot operations, no start corner is known as a rule. Therefore we use the dual simplex method to compute the first corner as is usually done in linear optimization.

A basic description is dual feasible if $c_N \geq 0$. We define the coefficients of the value function by 1 because the function can independently be chosen. If the considered simplex is placed sufficiently wide in the positive direction of the coordinate axes then the dual basic description of the coordinate origin has no redundant equations.

The change of the pivot element a_{rs}, that is computed by the dual pivot rule

$$p(s) = \max_{i \in B} \{\frac{b_i}{a_{is}} \mid a_{is} > 0, b_i < 0\}, \tag{3.7}$$

yields a new basic description that is dual feasible again. Analogously to the primal pivot operation with the quotient $q(s)$ and $\bar{q}(s)$, we derive a second quotient by means of the ε-transformation of the basic description mentioned above. Therefore let us consider the index set L

$$L = \{i \in B \mid \frac{b_i}{a_{is}} = p(s)\}$$

as the set of the indexes for which the attached quotient is equal to the characteristic quotient $q(s)$. Compute the characteristic quotient of the problem according to (3.7) changed by the ε-variation over the domain $B \cap L$. We get

$$t_2(s) = \max_{i \in B \cap L} \{-\frac{b_i + a_{iu}\varepsilon}{a_{is}} \mid a_{is} > 0 \ , \ b_i + a_{iu}\varepsilon < 0\} \ .$$

Analogously to (3.5), the dual pivot is defined by

$$\bar{p}(s) = \frac{1}{\varepsilon}(t_2(s) - p(s))$$

or

$$\bar{p}(s) = \min_{i \in B \cap L} \{\frac{a_{iu}}{a_{is}} \mid u \in N, u \neq s\} \ . \tag{3.8}$$

and all of the other restrictions described by L without of the restriction $r \in L$ are redundant ones.

Area Test of S

The algorithm mentioned above stops prematurely if

- the dual problem is not solvable ($S = \emptyset$),
- the primal step yields no new corner (S is a point),
- the second primal step yields no new corner (S is a straight line).

If we look for the corners of a straight line then it can be used the same scheme. The redundant equations have no influence on the computing time. One dual and one primal step yields the result and the answer whether the problem has not a solution, respectively.

The algorithm uses $O(n)$ operations to find a new corner according to (3.5), (3.6) and (3.7), (3.8) and to change the attached two variables in the simplex table. The greatest number of change steps is n because the algorithm takes, in each step, one index into N that did not hitherto belong to N. The total number of operation to enumerate all of the corners is $O(n^2)$.

If the simplex S has no redundancy restrictions at $\bar{x}$ itself, then only the restrictions of the coordinate system are redundant at the beginning of the dual steps. Therefore we get a corner after the second dual step. The area test takes only $O(n)$ operation in this case. Practical examples have shown, moreover, that an unfortunate location of the redundant half-spaces, which can essentially disturb the complexity $O(n)$, does not often arise.

The Transformation Algorithm

Linear systems of inequalities of the form (3.1) and a connection formula f can be transformed into their boundary representation in two steps:

- computation of a small as possible upperset of straight lines which covers all of the searched boundary, and
- characterization of the single straight line to be an edge of the modeled object.

The first step takes all of the half-space equalities, one behind the other, as the supporting set. For each supporting set, we ask for the intersection simplices (see Figure 3.10) of all convex objects that belong to the connection formula f.

The second step continues the algorithm. The resulting edges have to be split so that the relative inner points of each edge part belong to the same classification type. This is possible by pairwise intersecting all the considered edges. Lemma 2.18 gives the classification idea, and we can decide exactly which edge part belongs to which edge type or which edge part does not belong to any edge.

Without proof, the complexity of the transformation can be estimated in the following way. Let n be the total number of inequalities, let n_{max}

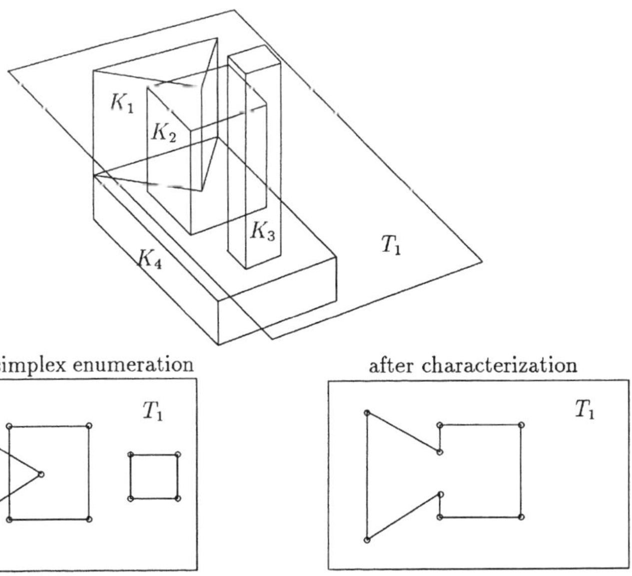

Figure 3.10. Boundary separation of $Q = K_1 \cup K_2 \cup K_3 \cup K_4$ in the supporting set T_1.

be an upper bound of each inequality number of a single object (system of inequalities), let m be the number of objects and the symbol $'$ signifies an evident reduction of the signed dimension symbol. The computation of the edges and their splitting can be estimated by $O(nn'_{max})$. The classification yields $O(nm''^2m + n^2_{max}m)$ operations. The tendency of these estimations can be shown in practice too.

4

A Small Language Modeler

The name GECO is taken from the first letters of the words 'GEometric COnstruction'. It is a PC modeling software for the geometric description of objects in three-dimensional Euclidean space E^3 by means of a special geometric language. It is possible to

- define one-, two- and three-dimensional objects,
- connect geometric objects by names (CSG-tree),
- compute boundaries of Boolean operations and convex hulls,
- store sequentially in terms of a hybrid representation and
- display geometric objects using hidden line algorithms.

The modeler consists of three parts, see Figure 4.1.

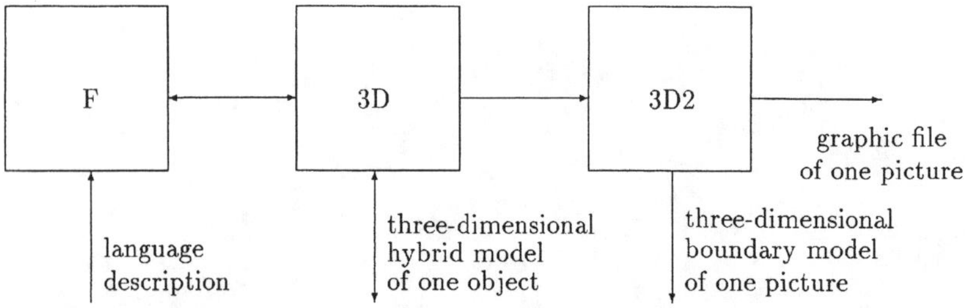

Figure 4.1. The three parts of the GECO modeler.

The language part F can interpret commands of the GECO dialog or it can interpret language files. It orders the geometric 3D part in order to build boundary and CSG-representations. The geometric 3D part organizes all operations with geometric objects. It can store and restore geometric objects in a hybrid description, i.e. boundary and CSG-representations together and sequentially. The passive viewer 3D2 takes only the internal description of the picture objects and is able to describe the three-dimensional boundary description or draw a picture on the screen. It can also write the polylines of a two-dimensional picture in a special file. In this way, we can build

- the language catalogue,
- the geometry catalogue and
- the graphic catalogue

by means of language files, geometry files and graphic files, respectively.

4.1. Principle of the Construction

It is not expected that any body can be described easily by a language text without a paricular systematic idea. The logic of GECO is based on the following modeling principle

- Well-defined elementary, geometric objects are created.
- Further geometric objects are constructed by set operations.

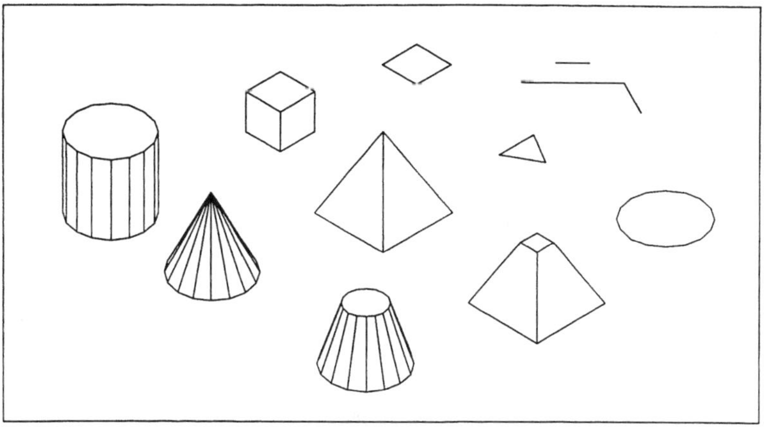

Figure 4.2. The elementary one-, two- and three-dimensional objects of the modeler.

The elementary objects are chosen in such a way that they are fixed only by a small number of parameters with a geometric reason. Each of the objects have only one name. So called *topologically elementary objects* are

- points, straightlines and planes.

There also exist *geometrically elementary objects* which form three classes, see Figure 4.2:

- 1-object: polyline
- 2-objects: rectangle, triangle, circle, convex polyhedron
- 3-objects: cuboid, pyramid, truncated pyramid, cylinder, cone, truncated cone, convex polyhedron.

Class operations connect operands of one and the same class:

- points: linear combinations
- i-objects: i-regular set operations.

In the case of points and 1-objects, the result is elementary again, the other cases are more complex. Connections between elements of different classes are

- the computation of convex hulls
- the connection by means of a common name.

The result of the convex hull computation leads to an elementary object again. Furthermore, it is possible to use integer and real numbers and their arithmetic for describing constructive parameters of the desired geometric object.

GECO is also an algorithmic language. Its language elements were chosen in such a way that all of the interesting connections of the given elementary objects can be expressed, and all of the geometric information about objects is reachable. In order to limit the size of the language, only one of all different possibilities for describing one and the same object is used.

Definition of Geometric Objects

The elementary geometric objects of GECO are defined by a command of the following form

$$DEF \quad A = SX;$$

where the variable A is defined as a geometric object of type S which is assigned the value described by X. One of the key words

POINT, *SRAIGHT*, *PLANE*,
PLAIN,
RECTANGLE, *TRIANGLE*, *CIRCLE*
CUBOID, *PYRAMID*, *PSTUMP*, *CYLINDER*, *CONE*,
CSTUMP, *POLY*

has to be used for the object type S. The description X of the corresponding elementary geometric object may consist of

three components (point),
a point list (polyline),
length and width (rectangle),
length, width and height (cuboid),
peak and base points (pyramid),
radius of the base and height (cone) and
an object list (convex polyhedron).

Objects also can be deleted by a corresponding command.

Operations with Geometric Objects

The essential operations with geometric objects are connections and transformations. The following connections are possible:

$A + B$:	union (or connection by name) of the objects A and B,
$A - B$:	difference, the result is the part of the object A that does not belong to B,
$A * B$:	intersection of A and B.

Furthermore, commands exist which can undo connections. In contrast to undoing connections by names or Boolean set operations, the computation of convex hulls are not reversible because the result is an elementary one again.

Let us connect the objects A, B and C with the result E:

$$TREE \quad E = A - B * (C + D) \ ;$$

this connection is at first a connection by the chosen name E. By computing the boundary representation of the object E,

$$BOUND \quad E \ ;$$

the new geometric data (vertices and edges) were computed. The "undoing" operation

$$BACK \quad E \ ;$$

breaks up the computed, geometric data into elementary ones. The most essential transformations of geometric objects are the locations. The translation of an object E by the vector F and the rotation of the object E on an axis G with the angle H are facilitated by the commands

$$MOVE \quad E \quad VECTOR \quad F \ ;$$
$$ROTATE \quad E \quad STRAIGHT \quad G \quad ANGLE \quad H \ ;$$

which move the points of geometric data. Furthermore, objects can be copied and elementary ones exchanged by other elementary ones in connected objects.

Input and Output of Geometric Information

The output of geometric objects happens in a three-dimensional form ($3D$) or as projections in two-dimensional form ($2D$). One example of a two-dimensional form is the projection

$CAVALIER$;
$COVERING$: $HALF$;
TO 1;

Parameters for the two-dimensional form specify the visibility and the lines necessary for approximating the actual geometric data. In the example, an orthogonal projection is chosen, lines of the surfaces at the back are deleted. All of the objects whose names are enumerated in the view name storage are drawn on the terminal with a cleared screen.

Analogously, numbers, three-dimensional objects in the form of files, protocol information and error messages can be written. If the output is a file, then geometric information can be stored and used in other programs as well as in the modeler.

Catalogues

Geometric descriptions such as

- programs in a geometric language,
- three-dimensional models or
- graphics

that are stored in any medium, may be combined and described in one document. The object selection of one catalogue has to correspond to the desired result. Catalogues can comprise information that is

- succinct and linguistically abstract,
- varied and easy comprehensible, three-dimensional and
- quickly retrievable graphic.

Program catalogues, 3D-model catalogues and graphic catalogues have to be distinguished in this sense. Graphic catalogues may be generated by 3D-model catalogues that may be themselves generated by program catalogues. Well-structured programs themselves document a program catalogue.

The complete geometric language of the GECO modeler can be found in the appendix. It is readable by means of the metalanguage described later in this chapter. The next page gives a symbolic description of the geometric language. This one page is very helpful when working in dialogue mode.

Therefore, the language examples that follow after this small language description of GECO can essentially be understood without studying the language description in depth.

A Short Language Description of GECO

files

GECO.EXE

GECO.DOK

.T text catalog

.G geometry catalog

program start

GECO

HELP;

ROW 1 MOP1 FROM 1;

END DIALOG;

protocol

PROT 1; error

PROT 7; modeling storage

··· until 7

OBINF A; object information

numbers

L = 5.7;

R = L + 7 * N;

MAX[I] = 1000;

MIN = −1000;

topological elements

DEF P = POINT (1, N + 1, 3.7);

DEF H = STRAIGHT P, (5, 3, 1);

DEF E = PLANE (5, 1), P, (8, 7, N);

1-object

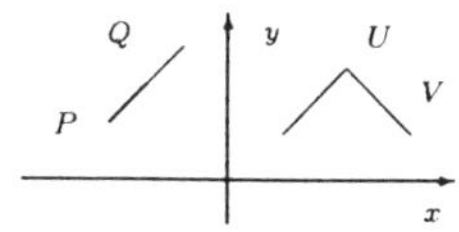

DEF S = PLINE P, Q, −(1, 1), U, V;

2-objects

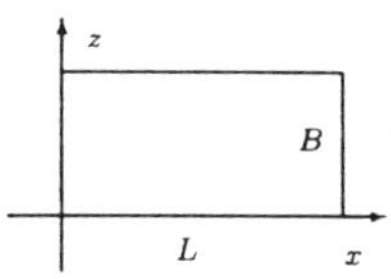

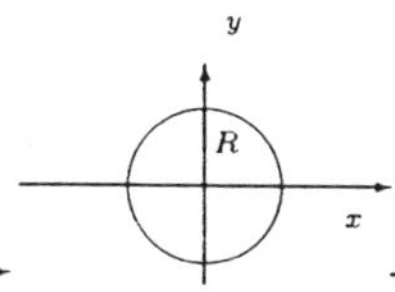

DEF C = CIRCLE R;

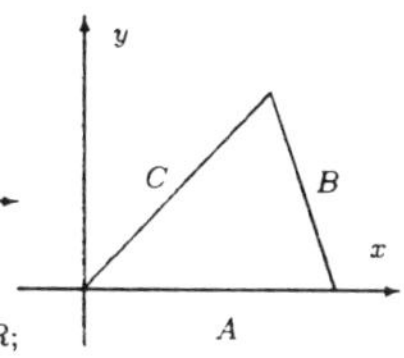

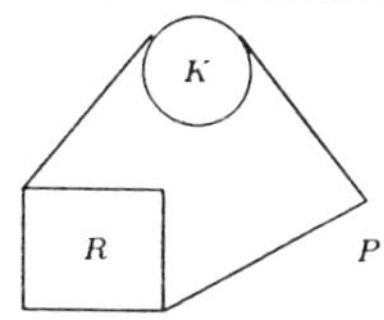

DEF P = POLY R, K, P;

DEF R = RECTANGLE L, B;

DEF T = TRIANGLE A, B, C;

3-objects

DEF Z = CYLI RADIUS R HIGHT H;

DEF U = POLY Q, Z, P, S;

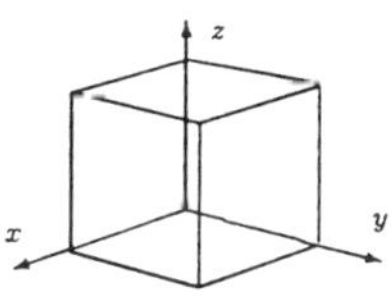

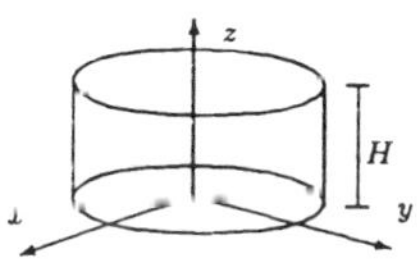

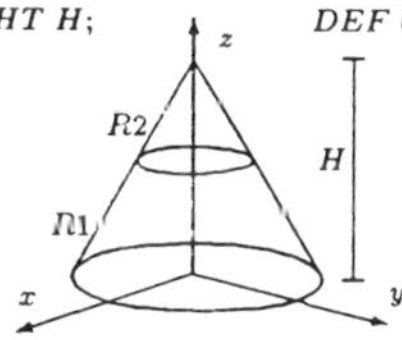

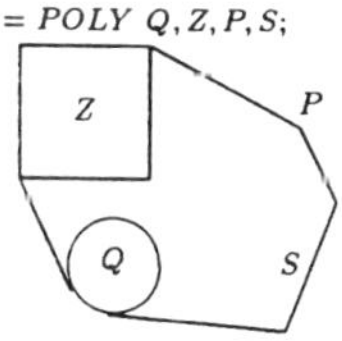

DEF Q = QUADER L, B, H;

DEF K = CONE RADIUS R1 HIGHT H;

DEF K2 = CSTUMP RADIUS R1, R2 HIGHT H;

localization

MOVE Q VECTOR (1, N + 1, 3.7);

ROTATE Q STRAIGHT G1 ANGLE 90G;

ROTATE Q STRAIGHT P1, P2 ANGLE 3, 14;

connection

TREE A = (B + C + D) − (U + V);

BOUND A;

DISSOLVE A;

BACK A;

picture

REDUCE ALL;	COVERING : NO;	ISOMETRY;
REDUCE A, B;	COVERING : HALF;	CAVALIER;
ADD A, B, C;	COVERING : COMPLETE;	DIMETRY;
TO 1;	APPROXIMATION : ALL;	PARALLEL ALPHA, BETA;
	APPROXIMATION : CONTOUR;	AXONOMETRY ALPHA, BETA, DX, DY, DZ;

file handling

DELETE ALL;	OUTPUT A EXT NAME AA TO 1;
DELETE A, B;	⟶ AA.G01
ROW 1 SBAG FROM 3;	INPUT A EXT NAME AA FROM 1;
SBAG.T03; ⟶	

4.2. Language Examples

Set Operations

Language file MOP.T01:

==

A simple example of the union

==

#DELETE ALL;	clearing of modeling storage

Tree connection between elementary objects

#DEF A = CUBOID 2, 1, 1;	definition of cuboids
#DEF B = CUBOID 1, 2, 1;	
#TREE C = A + B;	tree operation (connection by name)
#ADD C;	object C into drawing storage
#COVERING : NO;	kind of the covering
#CAVALIER;	projection
#TO 1;	drawing on screen

The new boundary

#BOUND C;	common new boundary representation
#COVERING : COMPLETE;	kind of covering
#TO 1;	drawing on screen
/	end of language file

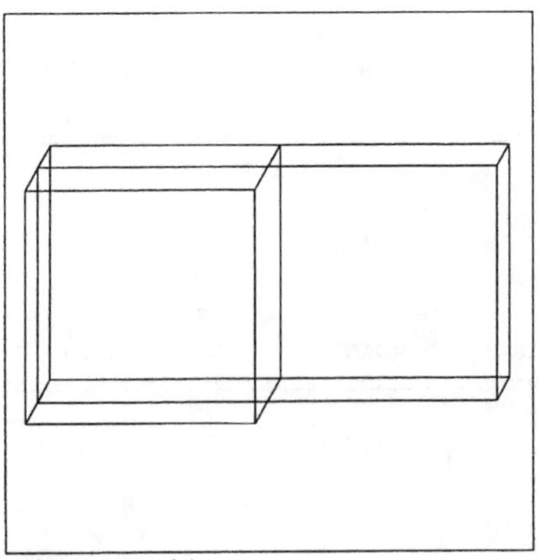

elementar objects

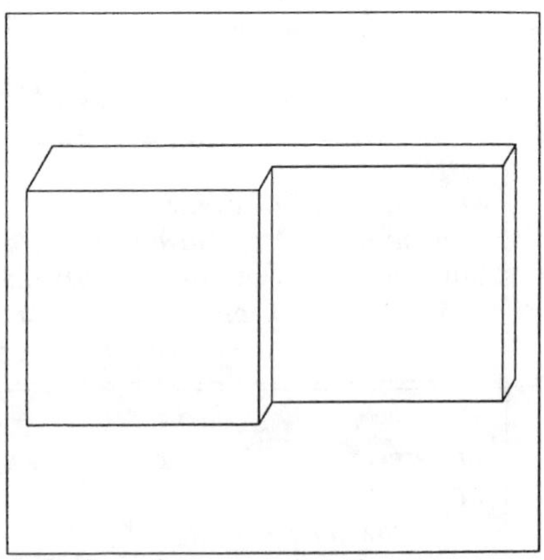

new boundary representation

Language file MOP2.T01:

```
  ==========================================
  A set operation with five cuboids
  ==========================================
#DELETE ALL;                        clearing of modeling storage
#CAVALIER;
#COVERING : NO;
  ------------------------------------
  Definition of the elementary objects
  ------------------------------------
#DEF A = CUBOID 5,5,1;
#DEF B = CUBOID 5,5,1;
#DEF C = CUBOID 5,1,5;
#DEF D = CUBOID 1,5,5;
#DEF E = CUBOID 1,1,1;
#DEF F = CUBOID 3,3,3;
#ADD A,B,C,D,E,F;                   names of the picture objects
#TO 1;                              draw on the screen
  ---------------------
  Positioning of objects
  ---------------------
#MOVE B VECTOR (0,0,2);
#MOVE C VECTOR (0,2,0);
#MOVE D VECTOR (2,0,0);
#MOVE E VECTOR (4,2,1);
#MOVE F VECTOR (1,1,1);
#COVERING : HALF;                   draw only forward faces
#TO1;                               draw on the screen
  -------------------
  A common CSG-tree
  -------------------
#TREE GGG = (A + B + C + D) - (E + F);
  ----------------------------------------------
  Computation of the new boundary representation
  ----------------------------------------------
#BOUND GGG;
#COVERING : COMPLETE;               draw without hidden lines
#REDUCE ALL;                        clear drawing storage
#ADD GGG;                           add to picture
#TO 1;                              draw on the screen
```

Exchanging of the projection

#*ISOMETRY*;	new projection
#*TO* 1;	draw on the screen
/	end of language file

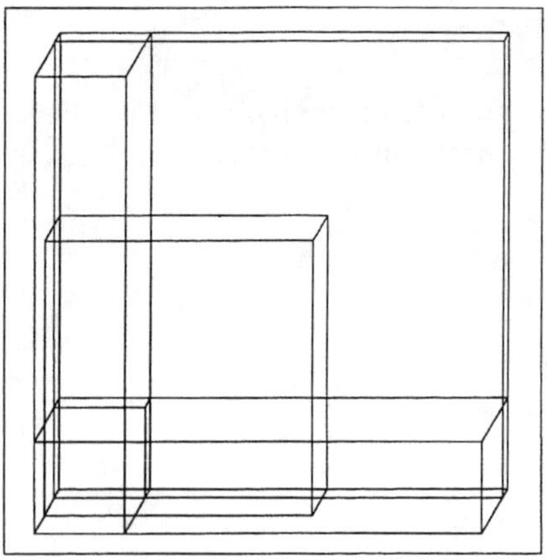

elementar objects

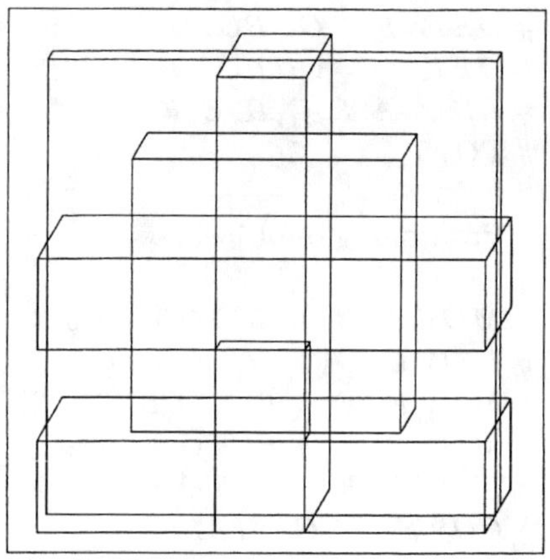

localization

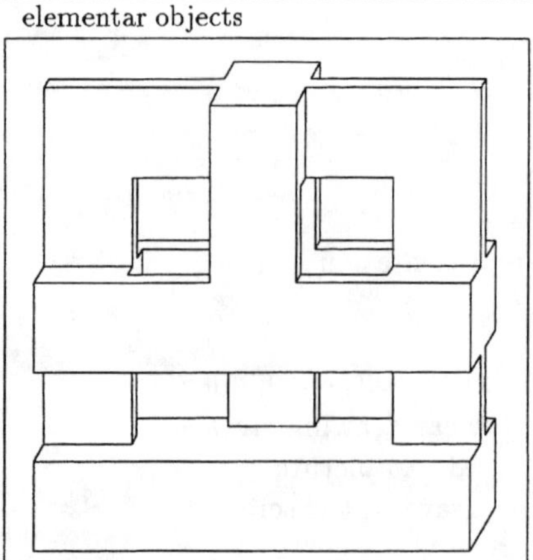

new boundary

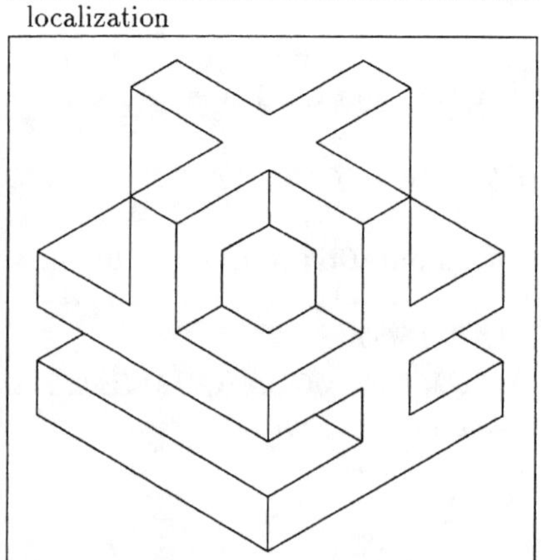

other projection

Convex Hull Construction

Language file KON1.T01:

```
==========================================
Construction of a stamp with convex hulls
==========================================
#DELETE ALL;                          clearing of modeling storage
#ISOMETRY;
#COVERING : HALF;
#APPROXIMATION : ALL;
-----------------------------------------
Definition and positioning of cylinders
-----------------------------------------
#DEF Z1 = CYLINDER RADIUS 1 HEIGHT 1;
#COPY Z2 = Z1;
#COPY Z3 = Z1;
#COPY Z4 = Z1;
#MOVE Z2 VECTOR (3,0,0);
#MOVE Z3 VECTOR (0,3,0);
#MOVE Z4 VECTOR (3,3,0);
#ADD Z2, Z3, Z4;
#TO 1;                                draw on the screen
-----------------------------------------
Definition and positioning of five faces
-----------------------------------------
#DEF R = RECTANGLE 2,2;
#DEF K1 = CIRCLE 0.7;
#DEF K2 = CIRCLE 1.5;
#COPY K3 = K2;
#COPY K4 = K1;
#MOVE R VECTOR (0.5,0.5,0.5);
#MOVE K1 VECTOR (1.5,1.5,6.0);
#MOVE K2 VECTOR (1.5,1.5,6.5);
#MOVE K3 VECTOR (1.5,1.5,7);
#MOVE K4 VECTOR (1.5,1.5,7.5);
#ADD R, K1, K2, K3, K4;
#TO 1;                                draw on the screen
--------------------------------
Drawing of three convex hulls
--------------------------------
#DEF U1 = POLY Z1, Z2, Z3, Z4, R;
#DEF P2 = POLY R, K1;
```

$\#DEF \;\; P3 = POLY \;\; K1, K2, K3, K4;$
$\#REDUCE \;\; ALL;$
$\#ADD \;\; U1, P2, P3;$
$\#TO \;\; 1;$ draw on the screen

Drawing again without approximation lines

$\#COVERING \; : \; COMPLETE;$
$\#APPROXIMATION \; : \; CONTOUR;$
$\#TO \;\; 1;$ draw on the screen
/ end of language file

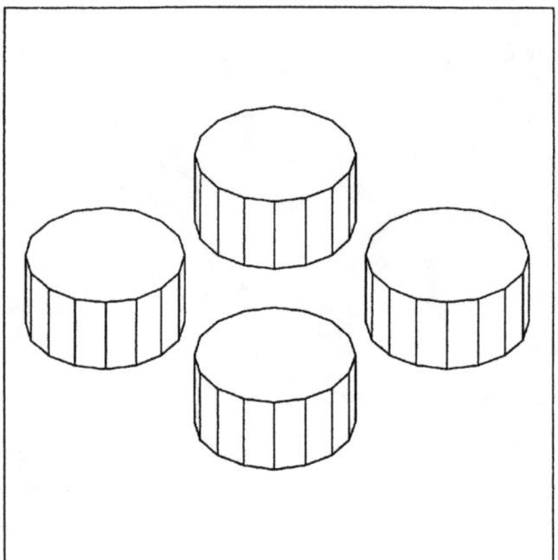

cylinders

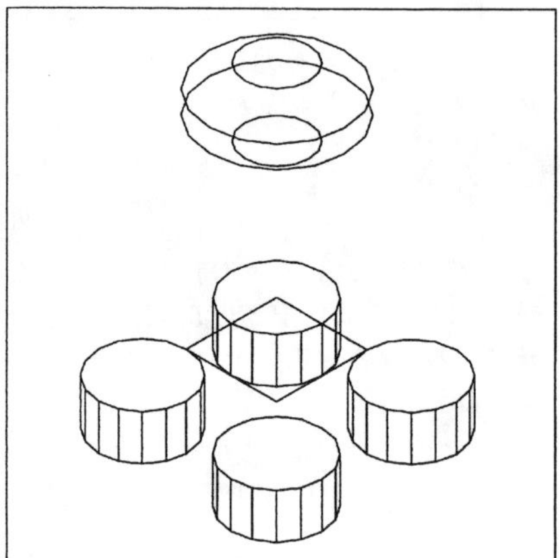

cylinder with faces

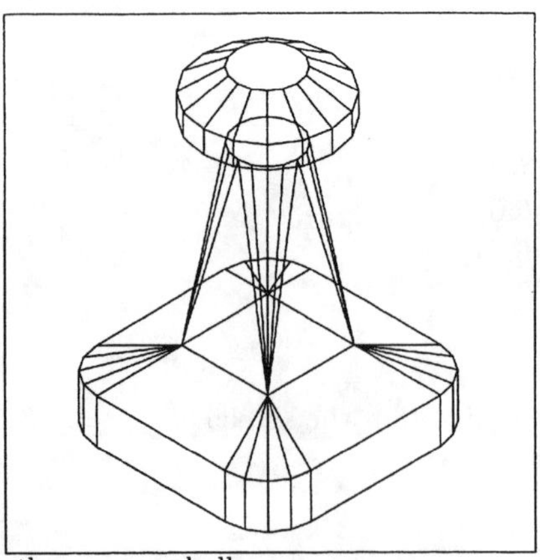

three convex hulls

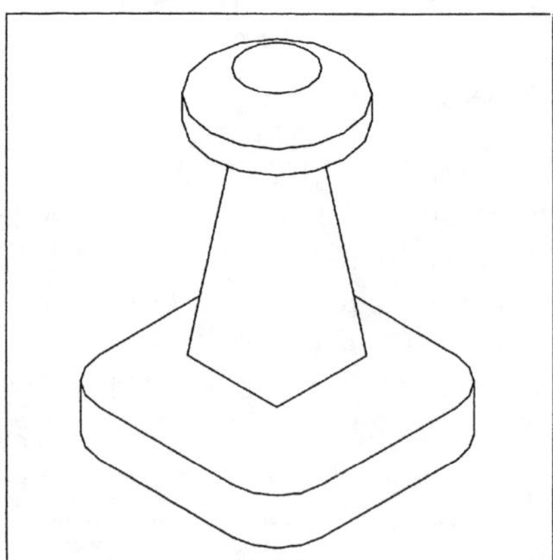

the stamp

Variable Construction with Parameters

Language file VAR1.T01:

```
=======================================
Description of a variable stamp part
=======================================
#DELETE ALL;                          clearing of modeling storage
-----------------------------
Parameter of the construction
-----------------------------
#REAL R = 1;                          rounding
#REAL B = 2;                          distance between the rounding
#REAL H = 1;                          height of a rounding
.................................choice...
Define R, B or H with other values and
continue on the written row number added by 1
...........................................
##                                    interruption of the language file
-----------------------------------------------
Definition and positioning of the elementary objects
-----------------------------------------------
#DEF Z1 = CYLINDER RADIUS R HEIGHT H;
#COPY Z2 = Z1;
#MOVE Z2 VECTOR (B,0);
#DEF ZZ1 = POLY Z1,Z2;
#COPY ZZ2 = ZZ1;
#MOVE ZZ2 VECTOR (0,B);
#DEF RE = RECTANGLE B,B;
#MOVE RE VECTOR (B/2,B/2,1.5*H);
#DEF PP = POLY ZZ1,ZZ2,RE;
-----------------------------------
Drawing of the elementary objects
-----------------------------------
#REDUCE ALL;
#ADD PP;
#COVERING : HALF;
#CAVALIER;
#TO 1;
/                                     end of language file
```

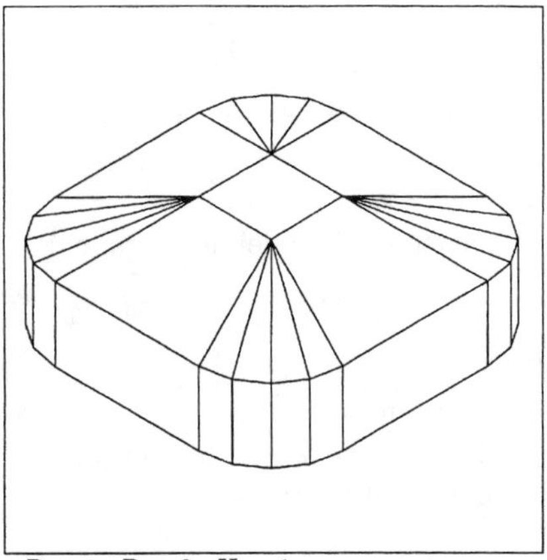

$R = 1$, $B = 2$, $H = 1$;

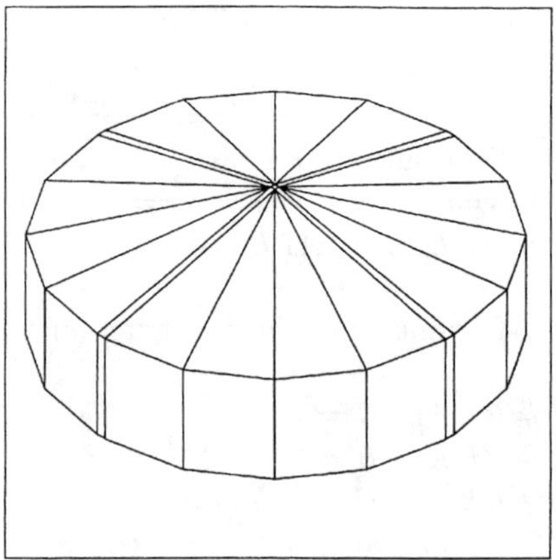

$R = 2$, $B = 0.1$, $H = 1$

Drawing

Language file *DAR2.T01*:

```
==========================================
48 cuboids with hidden
==========================================
#DELETE ALL;
#CAVALIER;
--------------------
Cuboids without covering
--------------------
#DEF Q1 = CUBOID 1,1,1;
#COVERING : NO;
#ADD Q1;
#TO 1;
------------------------------
Group of cuboids without backward faces
------------------------------
#COPY Q2 = Q1;
#COPY Q3 = Q1;
#MOVE Q2 VECTOR (1,0.5,0.5);
#MOVE Q3 VECTOR (2,0);
#REDUCE ALL;
#ADD Q1,Q2,Q3;
```

```
#COVERING : HALF;
#TO 1;
```

Copy and representation with half covering

```
#TREE Q4 = Q1 + Q2 + Q3;
#COPY Q5 = Q4;
#MOVE Q5 VECTOR (4,0);
#TREE Q6 = Q4 + Q5;
#COPY Q7 = Q6;
#MOVE Q7 VECTOR (0,3);
#TREE QQ1 = Q6 + Q7;
#COPY QQ2 = QQ1;
#MOVE QQ2 VECTOR (10,2);
#TREE QQ3 = QQ1 + QQ2;
#COPY QQ4 = QQ3;
#MOVE QQ4 VECTOR (20,0);
#TREE QUA1 = QQ3 + QQ4;
#REDUCE ALL;
#ADD QUA1;
#TO 1;
```

Drawing with complete covering

```
#COVERING : COMPLETE;
#TO 1;
/
```

end of language file

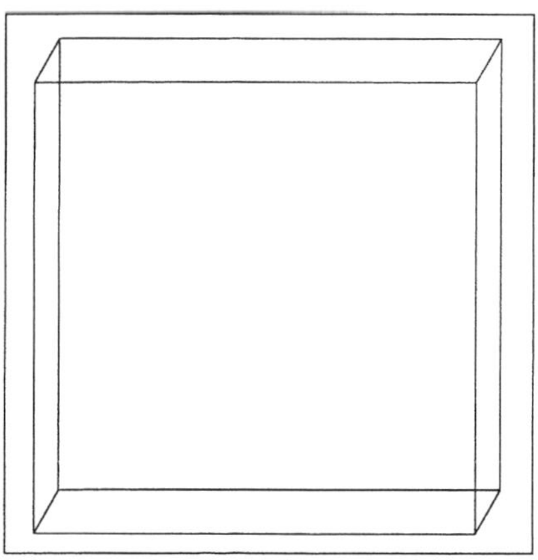

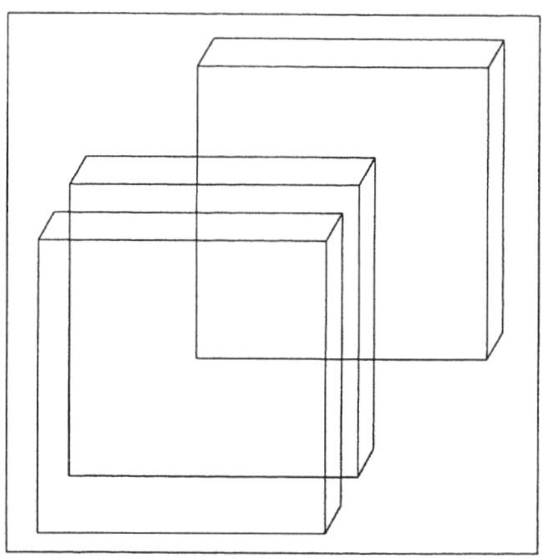

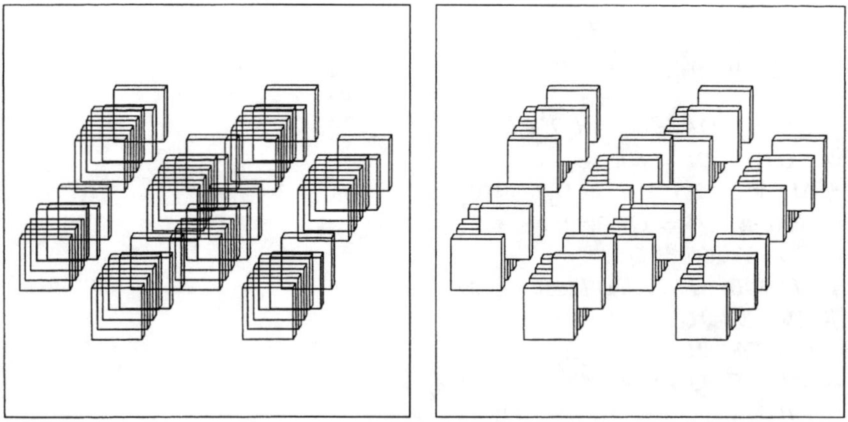

The following description of the metalanguage helps to understand the metalanguage itself (given in the appendix) and the description of the geometric files for storing the hybrid model (CSG and Boundary).

4.3. The Metalanguage

The syntax of the geometric language of the GECO modeler and its data interface will be written in the *metalanguage* code that consists of expansion of the well-known Backus Naur form.

Each syntax rule of a syntactic unit is given by the following form

non-terminal ::= right hand side

in order to define a new syntactic notion, the so called *non-terminals*, by means of *terminals*.

Terminals (key words of the language) are denoted by series of capital characters. The first four characters are significant. Non-terminals are series of characters too, but written in lower case. A word composition uses the symbol '_' to connect the words. A terminal cannot be used as a non-terminal. Non-terminals can be indicated by an index to reach a well-defined semantic description. An index is added to the character sequence as a partial word by the symbol '_'.

On the right hand side, alternatives, powers and lists as well as terminals and non-terminals may appear. Parenthesis are denoted by $< \cdots >$. *Language options*, for example between $a, b, ..., x$, arise in two forms:

$$a \mid b \mid \cdots \mid x \qquad or \qquad \left\{ \begin{array}{c} a \\ b \\ . \\ . \\ . \\ x \end{array} \right\}.$$

Language powers are for example

$$x < a : b > \qquad or \qquad < x >< a : b >$$

meaning that x is repeated at least a times but at most b times. It is possible that $a = 0$. Omitting b means that the upper bound is free. A *list* is denoted by the non-terminal

$$x\_list$$

and means that the non-terminal x arises at least once. Successive repetitions of x are possible. The repetitions are separated by the comma. In the metalanguage,

$$x\_list \quad ::= \quad x < \ , \ x >< 0 : >$$

defines the x_list.

4.4. Hybrid Geometric Files

Geometric object descriptions may be read from files and written into files for data transfer, data storage and cataloging.

The language files mentioned above are of the first kind. The description consists of geometric instructions to construct the desired geometric object.

The hybrid files are the second kind of geometric description. They can store and restore the common CSG and boundary description of an object in terms of the *r_object* (real number), *i_object* (integer number), *po_object* (point), *s_object* (straight), *p_object* (plane), *l_object* (polyline), *f_object* (face), *b_object* (body) and the *tree_object* (combination of the enumerated objects).

Some Elementary Lists Describing Objects

The following lists are used in the desired object descriptions.

$$\begin{array}{lcl} po\_list & ::= & point < 1 : > \\ point & ::= & number \end{array}$$

Points ($POINTS$) and *point lists* (po_list) help to represent vertices of geometric objects.

$$\begin{array}{lcl} l\_list & ::= & line < 1 : > \\ line & ::= & < \;\; - \;\; integer < 2 : > >< 1 : > \end{array}$$

Polylines are represented by means of integer sequences in terms of the *line list* (l_list) that denote the pointers to the associated point in the point list. Two integers following each other define the point numbers of a single line if both of them are positive. Otherwise they define a non-existing line (not visible).

$$f\_list \;\; ::= \;\; < \;\; l\_list \cdot integer\_3 \;\; >< 1 : >$$

Face lists (f_list) are line list with closed polylines. The symbol $integer\_3$ defines the pointer of the beginning of the closed polyline which describes a boundary representation of a single face.

$$\begin{array}{lcl} b\_list & ::= & head \;\; f\_list < 2 : > \\ head & ::= & integer\_1 \;\; < \;\; integer\_2 \;\; integer\_3 \;\; >< 0 : > \end{array}$$

several face lists can be combined to a *body list.*

$integer\_1$: number of faces
$integer\_2$: enumeration number of a face
$integer\_3$: number of points in the respective face list.

$$n\_list \;\; ::=$$
$$objectname\_1 < objectname\_2 \left\{ \; - \; \right\} < 1 : >> \left\{ \begin{array}{c} + \\ - \end{array} \right\} < 1 : 0 >$$

The *notation list* (n_list) connects the objects by means of the Polish notation.

Input and Output Lists of Auxiliary Objects

$$\begin{array}{lll} r\_object & ::= & REAL\ \ name\ \ number\ \ TEXT < 0:1 > \\ i\_object & ::= & INTEGER\ \ name\ \ integer\ \ TEXT < 0:1 > \end{array}$$

Real and integer objects (*r_object* and *i_object*) help to define parameters of geometric objects. They are named numbers.

$$\begin{array}{lll} po\_object & ::= & POINT\ \ name\ \ point\ \ text < 0:1 > \\ s\_object & ::= & STRAIGHT\ \ name\ \ point\ \ point\ \ text < 0:1 > \\ p\_object & ::= & PLANE\ \ name\ \ point\ \ point\ \ point\ \ text < 0:1 > \end{array}$$

Just as number objects, the object lists *po_object* (point list), *s_object* (straight list) and *p_object* (plane list) of non-restricted geometric objects are defined.

Input and Output Lists of Geometric Objects

$$\begin{array}{lll} l\_object & ::= & LINIE\ \ name\ \ text < 0:1 > \\ & & PL\ \ integer\ \ p\_list\ \ LT\ \ integer\ \ l\_list \\ f\_object & ::= & FACE\ \ name\ \ text < 0:1 > \\ & & PL\ \ integer\ \ p\_list\ \ LT\ \ integer\ \ f\_list \\ b\_object & ::= & BODY\ \ name\ \ text < 0:1 > \\ & & PL\ \ integer\ \ p\_list\ \ FL\ \ integer\ \ b\_list \end{array}$$

Here, PL means *point list*, LT *line list* and FL *face list*. Polylines (*l_object*), faces (*f_object*) and bodies (*b_object*) are characterized by point lists and connections (line lists, face lists and body lists). The term *integer* tells the size (number of the points and number of the pointers to the point list, respectively) of the list.

$$\begin{array}{lll} tree\_object & ::= & object\ \ name \\ & & \quad t\_object < 0: > \\ & & \quad g\_object < 0: > \\ & & \\ t\_object & ::= & type\ \ name\ \ text < 0:1 > \\ & & \quad NL\ \ integer\ \ n\_list \\ & & \quad < PL\ \ integer\ \ p\_list\ \ t\_list >< 0:1 > \end{array}$$

(Continued)

$$g\_object \quad ::= \quad l\_object \mid f\_object \mid b\_object$$

$$type \quad ::= \quad LINIE \mid FACE \mid BODY$$

$$t\_list \quad ::= \quad \left\{ \begin{array}{l} LT \;\; integer \;\; l\_list \\ LT \;\; integer \;\; f\_list \\ FL \;\; integer \;\; b\_list \end{array} \right\}$$

NL means *notation list.* The constructive relation of a complicated object to an elementary one is reflected by the *tree list* (t_object). If a *geometric list* (g_object) follows the notation list of a tree object then its boundary representation is computed and follows directly with all geometric information of its boundary representation. If the elementary objects of the tree object are faces or bodies, then they have to be convex ones. The term $integer$ defines the length of the following list.

Example:

The following language description of an object is given (see Figure 4.3).

$$\begin{array}{l} DEF \;\; A = CUBOID \;\; 2,1,1; \\ DEF \;\; B = CUBOID \;\; 1,2,1; \\ DEF \;\; C = CYLINDER \;\; RADIUS \;\; 0.5 \;\; HEIGHT \;\; 1; \\ MOVE \;\; C \;\; VECTOR \;\; (0.5, 0.5, 1); \end{array}$$

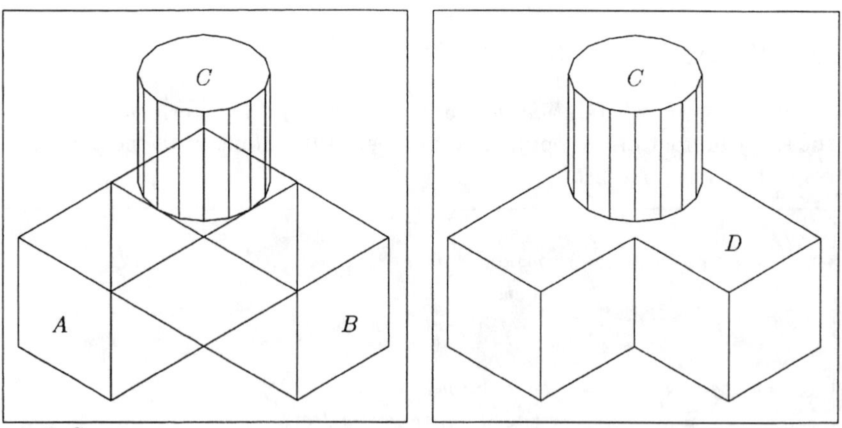

A, B, C $\qquad$ $E = C + D$

Figure 4.3. Connection of two cuboids with a cylinder.

```
TREE D = A + B;
BOUND D;
TREE E = C + D;

OUTPUT E EXT NAME EMIL TO2;

ISOMETRY;
COVERING : COMPLETE;
ADD E;
TO 1;
```

It produces the following file *EMIL.G02* in ASCII format in the form mentioned above.

```
OBJECT  EMIL
BODY    EM05   TREE
NL      3
        EM03   EM04    +
BODY    EM04   BOUN
NL      3
        EM02   EM01    +
PL      12             (number of points)
1.000  2.000  0.000
1.000  1.000  0.000
0.000  2.000  0.000
0.000  0.000  0.000
 ...    ...    ...
 ...    ...    ...
 ...    ...    ...
2.000  1.000  1.000
FL  61               (length of the list in words)
8    1   7  2   7   3    5    4   5   5
5    6   5  7   5   8    5   -1   2   6
5    4   3  1  -7   8    9   10  11  12
7  -10   9  3   4  10  -11   10   4   5
11  -8   7  2   1   8   -9    8   1   3
9  -12  11  5   6  12   -7   12   6   2
7
BODY  EM03  ELEM
PL    32
0.962  0.309  1.000
0.854  0.146  1.000
0.691  0.038  1.000
```

```
0.500  0.000  1.000
0.309  0.038  1.000
0.146  0.146  1.000
0.038  0.039  1.000
0.000  0.500  1.000
0.038  0.691  1.000
0.146  0.854  1.000
0.309  0.962  1.000
0.500  1.000  1.000
0.691  0.962  1.000
0.854  0.854  1.000
0.926  0.691  1.000
1.000  0.500  1.000
0.962  0.309  2.000
 ...    ...    ...
 ...    ...    ...
 ...    ...    ...
1.000  0.500  2.000
FL  151
18   1    17  2   17   3   5    4   5   5
5    6    5   7   5    8   5    9   5   10
5    11   5   12  5    13  5    14  5   15
5    16   5   17  5    18  5    -1  2   3
4    5    6   7   8    9   10   11  12  13
14   15   16  1   -32  31  30   29  28  27
26   25   24  23  22   21  20   19  18  17
32   -1   17  18  2    1   -2   18  19  3
2    -3   19  20  4    3   -4   20  21  5
4    -5   21  22  6    5   -6   22  23  7
6    -7   23  24  8    7   -8   24  25  9
8    -9   25  26  10   9   -10  26  27  11
10   -11  27  28  12   11  -12  28  29  13
12   -13  29  30  14   13  -14  30  31  15
14   -15  31  32  16   15  -16  32  17  1
16
BODY  EM02  ELEM
PL    8
0.000  0.000  0.000
2.000  0.000  0.000
2.000  1.000  0.000
0.000  1.000  0.000
0.000  0.000  1.000
2.000  0.000  1.000
```

```
2.000  1.000  1.000
0.000  1.000  1.000
FL  43
6  1  5   2  5  3  5  4   5  5
5  6  5  -1  4  3  2  1  -5  6
7  8  5  -1  5  8  4  1  -1  2
6  5  1  -2  3  7  6  2  -7  3
4  8  7
BODY  EM01  ELEM
PL    8
0.000  0.000  0.000
1.000  0.000  0.000
1.000  2.000  0.000
0.000  2.000  0.000
0.000  0.000  1.000
1.000  0.000  1.000
1.000  2.000  1.000
0.000  2.000  1.000
FL  43
6  1  5   2  5  3  5  4   5  5
5  6  5  -1  4  3  2  1  -5  6
7  8  5  -1  5  8  4  1  -1  2
6  5  1  -2  3  7  6  2  -7  3
4  8  7
```

The external name $'EMIL'$ was chosen for this example. This name arises as the object name. The names of the elementary parts and partial objects are connected by the name $'EMIL'$. It yields the new name from the first two letters and the first letter, respectively, of the external names and in continuation from a number to complete the name to four characters.

4.5. General Remarks

When the GECO program is used, special conditions have to be taken into account, i.e., installation conditions, error handling and some restrictions of a geometrical or numerical kind.

Installation Conditions

The system needs a PC with MS-DOS and co-processor because the set operation are time consuming. A VGA is used for drawing.

On the server (see demo address, p. viii) provided, GECO contains all information needed to start the system and demonstrate some small applications. Information about the system and the executable program is stored in the following files:

hints		readme.txt
executable program		GECO.EXE
language description		IDIOMS.DOK
catalog 1		
	elementary objects	GREL.T01
	simple set operations	MOP1.T01
	trees, geometric file	MOP2.T01
	convex hulls	KON1.T01
	parameter construction	VAR1.T01
catalog 2		
	an optical illusion	DAR1.T02
	maximum of hidden lines	DAR2.T02
	DO-instruction	KUGE.T02
	a hybrid model	CUT3.T02
		EMIL.G02
catalog 3		
	a large model	SBAG.T03

It is advantageous to transfer the files to the disk, and even better to the RAM-disk. After starting, the system prompts with a pound sign.

If you have forgotten any of the language instructions, then you have only to write

$$\begin{array}{ll} \#HELP; & <CR> \\ \# & <CR> \end{array}$$

to get information about suitable commands. Two kinds of communication are possible. Firstly, the beginning of the language files can be written with a standard editor. Secondly, we can write dialog instructions as mentioned above.

A language file $'name.T01'$ is begun with the first row by

$$\begin{array}{ll} \#ROW\ \ 1\ \ name\ \ FROM\ \ 1; & <CR> \\ \# & <CR> \end{array}$$

where '*name*' has up to four digits. Each row of the language file will be shown on the screen and you can control the modeling process. The rows are continued by the comment

$$GECO\ \ name\ \ rownumber$$

to show the actual instruction. If the process stops before the end was detected, then you can use the dialog to correct the wanted instruction and you can continue with the current row number to finish the process. It may be a good idea to get started using the demonstration files $' * .T01'$.

In the dialog, we write some instructions and end each instruction by $< CR >$. They are put into a small *instruction storage.* If we quit one of the pound signs by $< CR >$ without writing any new instruction, then the prior instructions are checked for syntax errors and processed. It is better in the dialog to process only one instruction to detect typing errors. Only *DO* instructions and conditions need the small instruction storage.

Remarks:

1. If the last session was finished with something other than the break $< CTRL \;\; C >$, then the new session begins with the first modeling storage from the beginning of the last session. Otherwise, if it was finished with the command

   $$END \;\; DIALOG;$$

   then the result of the last session is taken as the beginning of the new session. After that, the following instructions

   $$PROT \;\; 2,$$
   $$OBINF \;\; name;$$

   are suitable to recognize which of the names are chosen ($PROT$), which of objects exist and prehaps which geometric type ($OBINF$) they are.

2. The modeling storage is named by $'MODSP.DAT'$. Some temporary files ($' * .DAT'$) are produced to draw a picture. They can be deleted at any time.

Error Handling

The system accepts a syntactically correct instruction from a language file if the text $'GECO \;\; name \;\; rownumber'$ on the end of the corresponding row appears. Otherwise, an error message appears or the text $'GECO \;\; ERROR \;\; rownumber'$ is displayed and processing stops at which point control is back to the dialog. Dialog errors are handled analogously. The instruction

$$PROT \;\; 1;$$

gives some information about the kind of error. The language file can be continued by starting with a higher row number.
A list of all possible error messages and some hints are given in the appendix. It is possible to formulate examples that are geometrically unstable with the given geometric instructions of the special language. The programming language used was F77, with its insufficient number of digits of the mantissa. From a theoretical position, it is not possible to program stable set operation using only double precision in FORTRAN (see later). Therefore, GECO is a compromise between numerical stability and the portable programming using only FORTRAN.

Examples can be given to show that convex hulls sometimes disturb the Boolean set operation. Then the idea of the construction has to be changed. If the process stops during computation of such an unstable set operation and modeling is to be continued, then it is advisable to start the used language file with a row after the instruction $BOUND \cdots$.

Error Messages of the Modeler

The small language modeler GECO recognizes the following errors and gives a brief explanation in each case to correct the description.

object is already defined	- take another name
missing object	- define an object
missing of an elementary part	- give the right name
missing subprogram	- exchange the wrong word
wrong object type	- exchange the type
wrong device	- exchange the device
list overflow	- exchange the number of objects, points, edges or points per face
arithmetic overflow	- change the number domain
geometrically impossible	- change the construction
point coordinates overflow	- domain -1000...1000
overflow of the name or variable number	- delete names or variables
syntactically wrong names or variables	- exchange names or variables
syntactically wrong index of a component	- exchange index
syntactically wrong index of a variable	- exchange variable
indicated name too long	- take shorter name
stack overflow of arithmetic terms	- take a smaller term
computation not allowed	- control the used functions

wrong arithmetic term	- control the term syntax
error in the standard function	- exchange standard function
wrong real number	- exchange real number
integer too large	- take a smaller integer
missed semicolon	- write a semicolon
comment not recognizable	- signify the comment
too many parameters	- take fewer parameters
comment error	- control the comment
syntax error	- correct the syntax
term syntactically wrong	- vector or set operation
too large notation or stack	- take a smaller expression
wrong character in the term	- exchange this character
wrong logical term	- correct the term
wrong condition	- correct the condition
wrong *DO*-instruction	- correct the *DO*-instruction
no end of an instruction	- write an *END*
wrong end of an instruction	- correct the end
ELSE statement wrong placed	- place it right
parameter list syntactically wrong	- correct the syntax

Restrictions of this Version

1. Calls of language files, conditions and DO-instructions cannot call themselves.

2. In the dialog, a maximum number of 16 rows at the same time are possible. One can process the instructions one after another. The restriction has to be noticed in conditions and DO- instructions. Their row number should be smaller than 17.

3. The graphic files cannot be internally named. Therefore they have to be named with a catalog name by means of the operating system MS-DOS.

4. Using Boolean set operations ($BOUND \cdots$), the number of points and edges of one face is restricted. It is possible to compute faces with up to 300 points and 1000 edges.

5. The modeling storage 'MODSP.DAT' contains up to 1500 points, 6000 edges and 300 object names.

4.6. Key Words and Characters

The following key words are used in the language GECO. Please note that the single characters D and E are key words themselves and they cannot be taken as a name of any variable.

Information:	*PROT, OBINF, HELP*
Terms:	*REAL, INT, D, E* *EXP, LOG, TAN, SIN, SQRT, COS,* *COT, ABS, ATAN, ASIN, ACOS, ACOT*
Relations:	*EQ, NE, LT, GT, LE, GE, NOT, AND, OR*
Descriptions:	*DEF, DELETE, COPY, TREE, BOUND,* *SUBSTITUTE, BACK, DISSOLVE, FROM,* *DETAIL, INVERT*
Auxilary objects:	*POINT, STRAIGHT, FACE, INTERSECT,* *EQUATION*
Objects:	*PLINE, LINE,* *RECTANGLE, TRIANGLE, CIRCLE, FACE,* *CUBOID, CYLINDER, CONE, CSTUMP,* *PYRAMID, PSTUMP, POLY,* *GROUNDF, HEIGHT, RADIUS*
Transformations:	*MOVE, VECTOR, ROTATE, ANGLE,* *ORIENTATION, INVERT*
Drawing:	*FRONTVIEW, SIDEVIEW, TOPVIEW,* *ISOMETRY, DIMETRY, CAVALIER,* *ORTHOGONAL, AXONOMETRY*
Drawing:	*COVERING, NO, HALF, COMPLETE,* *APPROXIMATION, CONTOUR, ALL,* *ADD, REDUCE*
Storage:	*CANAL, OUTPUT, INPUT, EXT, NAME, TO,* *FROM*
Controls:	*IF, THEN, ELSE, WHILE, DO, ROW, DIALOG,* *END*

A , B , C , D , E , F , G , H , I , J , K , L , M
N , O , P , Q , R , S , T , U , V , W , X , Y , Z

0 , 1 , 2 , 3 , 4 , 5 , 6 , 7 , 8 , 9

, , . , = , (,) , ' , + , - , * , / , # ,   , ;

a , b , c , d , e , f , g , h , i , j , k , l , m
n, o , p , q , r , s , t , u , v , w , x , y , z

5

The Algebraic Model

This chapter deals with the definition of abstract algebras for modeling. Boolean algebra can be used as an algebraic model and moreover, we need this algebra for a basic description of a more abstract model.

5.1. The Boolean and the Weak Boolean Algebra

The following definition of a Boolean algebra is very old but it is practical for modeling nowadays. We only collect the well-known assertions in the following section, most of them without any proof. More about Boolean algebras can be found in special algebraic books like [Ku 68] and in those geared towards applications like [Wh 04].

The Boolean Algebra

A mapping f from the Cartesian product $B \times B$ into B is called a *binary operation*. The function f assigns the two *operands* $x, y \in B$ to the *result* $z \in B$. A *single operation* or *multiple operations*, can be represented by binary ones.

Definition 5.1. *A fundamental set B and two binary operations are called a* **Boolean algebra** $(B, +, *)$ *if and only if the following axioms hold:*

B1 *The operations* $(+)$ *and* $(*)$ *are commutative ones.*

B2 *There exists two unit elements,* 0 *and* 1*, of the operations* $(+)$ *and* $(*)$*, respectively.*

B3 *Each operation is distributive to the other.*

B4 *In each operation, there exists an inverse element* $a' \in B$ *for all* $a \in B$*.*

The element a' which is assigned to the element a in a Boolean algebra B is always well-defined. It is called the *complement* of a. The operation $(+)$ is called *addition* and the operation $(*)$ is called *multiplication*. For the simplification of the notation, we write only ab instead of $a * b$. According to the definition, the Boolean algebra is closed with respect to its operations, i.e., the results from applying each of the operations to elements in B, as well as to the complement of elements in B, are again in B. Multiple operations of the Boolean algebra are interpreted using the following priority rule for reducing to binary operations or to the complement:

complement before multiplication before addition.

The priority rule can be modified by including subterms in parentheses.

A *constant* is a symbol that represents a fixed element of the Boolean algebra, or a term that can be represented by a fixed element. The elements 0 and 1 are examples of constants. A *variable* is a symbol (e.g., x or y) which represents a non-fixed element of the Boolean algebra. Any multiple operation constructed from a finite number of variables and constants, by means of the operations $(+)$ and $(*)$ as well as their complements, is called a *Boolean function*. For example, let a and c be constants and x and y be variables. Thus the term $((a' + 0)'y + ac')x$ represents a Boolean function. Two Boolean functions f and g are called a *disjunctive function* if the product of their results is equal to 0.

In order to describe Boolean functions unambiguously and to provide a survey of all possible Boolean functions, we need the following notions and assertions. We are particularly interested in those functions of n variables $x_1, x_2, \cdots, x_n$, which we can write additively and whose summands are products of all variables with or without the complement symbol. Examples of those functions of one, two and three variables are $x + x'$, xy' and $xyz' + x'yz + xy'z$. The following definition gives a name to those functions.

Definition 5.2. *A Boolean function is given in* **disjunctive normal form** *in* n *variables* $(n > 0)$*, if its term is a sum over terms of the form* $f_1(x_1)f_2(x_2)\cdots f_n(x_n)$*, in which* $f_i(x_i)$ *is either* x_i *or* x'_i *for each* i $(i = 1, 2, \cdots, n)$*, and if no two terms coincide in the sum.*

Simple constants are always given in a disjunctive normal form of n variables. Some fundamental properties of the disjunctive normal form are explained in the following statements.

Lemma 5.3. *Each function in the Boolean algebra which does not contain a constant is equal to a function in disjunctive normal form.*

Each function can be represented in arbitrarily many different normal forms by changing the number of variables. For example the function $f = xy$ is given in normal form with respect to x and y. Multiplying f by $(z + z')$ we get $f = xyz + xyz'$. Now f depends on the three variables x, y and z, and it is given in normal form. From this point, we shall assume that the notation of the disjunctive normal form is related to that normal form which has the minimal number of variables.

With this restriction, it can be shown that each function defines exactly one normal form with a given number of variables, as is to be seen in the assertions below.

The single summands of the normal form are disjunctive to each other and dependent on each variable x_i $(i = 1, 2, \cdots, n)$. Now we choose only one summand from the set of all possible summands of a disjunctive normal form in n variables. This corresponds to the selection of either x_i or x_i' for each of the n variables x_i $(i = 1, 2, \cdots, n)$. Therefore, there are maximal 2^n different summands, which can be found in a disjunctive normal form of n variables. This disjunctive normal form, consisting of all 2^n summands is called the *complete disjunctive normal form* in n variables. The complete form will be shown to be identically equal to 1.

Lemma 5.4. *If one of the values 0 or 1 is assigned to each of the n variables, in any but a fixed manner, then exactly one summand of the complete disjunctive normal form in n variables has the value* 1*, while all the others have the value* 0.

Proof. Let $a_1, a_2, \cdots, a_n$ be the values of $x_1, x_2, \cdots x_n$, and each a_i is 0 or 1. We choose a summand of the complete disjunctive normal form, referring to the following rule:

If $a_i = 1$ then x_i must belong to the summand to be chosen. On the other hand, if $a_i = 0$ then x_i' must belong to the summand for each i $(i = 1, 2, \cdots, n)$.

This leads to a chosen summand which is a product of n factors equal to 1 and so it is equal to 1. On the other hand, all the other summands of the complete disjunctive normal form contain at least one factor equal to 0 and therefore they are equal to 0. □

Lemma 5.5. *Two functions are equal to each other if and only if their disjunctive normal forms contain the same summands.*

In order to obtain or verify an identity in Boolean algebra, it is sufficient to examine the values of each function for all combinations of 0 and 1 which can be put in the variables.

In this sense, each Boolean function can easily be described by establishing a value table for all combinations with the symbols 0 and 1, instead of the variables themselves. If such a table is given, then the function in disjunctive normal form can be obtained with that table. For each combination of the values 0 and 1, for which the function shall have the value 1, the adequate summand of the disjunctive normal form is added. This was indicated in the proof of Lemma 5.4. The sum over these summands yields the function.

We denote the 2^n summands of the complete disjunctive normal form in n variables $x_1, x_2, \cdots, x_n$ as $g_1, g_2, \cdots, g_m$ with $m = 2^n$. Now we choose only one disjunctive normal form from the set of all possible disjunctive normal forms of the n variables. This implies that the disjunctive normal form to be chosen either contains g_i or not g_i for each $i(i = 1, 2, \cdots, n)$. Therefore there are exactly 2^m with $m = 2^n$ disjunctive normal forms of the n variables $x_1, x_2, \cdots, x_n$.

Table 5.1 gives the 16 possible binary operations (Boolean functions). The so called difference $a * b'$ has the symbol $a - b$ and the so called symmetric difference $(a - b) + (b - a)$ has the symbol $a \bigtriangledown b$.

Operations can be demonstrated by means of the well-known *Euler Venn diagram.* In this diagram, the operands are represented as areas overlapping each other. The parts of the areas belonging to the result are outlined. The boundaries of the operand area divide the whole area into disjunctive parts of inner area points. Each disjunctive part corresponds to a summand of the disjunctive normal form in a Boolean algebra. The sum over all such inner points represents the unit element 1, and is drawn with a rectangle. Table 5.2 shows the 16 possible Boolean operations.

$H1(a,b)$	$=$	$a + b$	$H9(a,b)$	$=$	$(a+b)'$
$H2(a,b)$	$=$	$a * b$	$H10(a,b)$	$=$	$(a*b)'$
$H3(a,b)$	$=$	$a - b$	$H11(a,b)$	$=$	$(a-b)'$
$H4(a,b)$	$=$	$b - a$	$H12(a,b)$	$=$	$(b-a)'$
$H5(a,b)$	$=$	$a \bigtriangledown b$	$H13(a,b)$	$=$	$(a \bigtriangledown b)'$
$H6(a,b)$	$=$	a	$H14(a,b)$	$=$	a'
$H7(a,b)$	$=$	b	$H15(a,b)$	$=$	b'
$H8(a,b)$	$=$	0	$H16(a,b)$	$=$	1

Table 5.1. Listing of the 16 possible binary operations in a Boolean algebra.

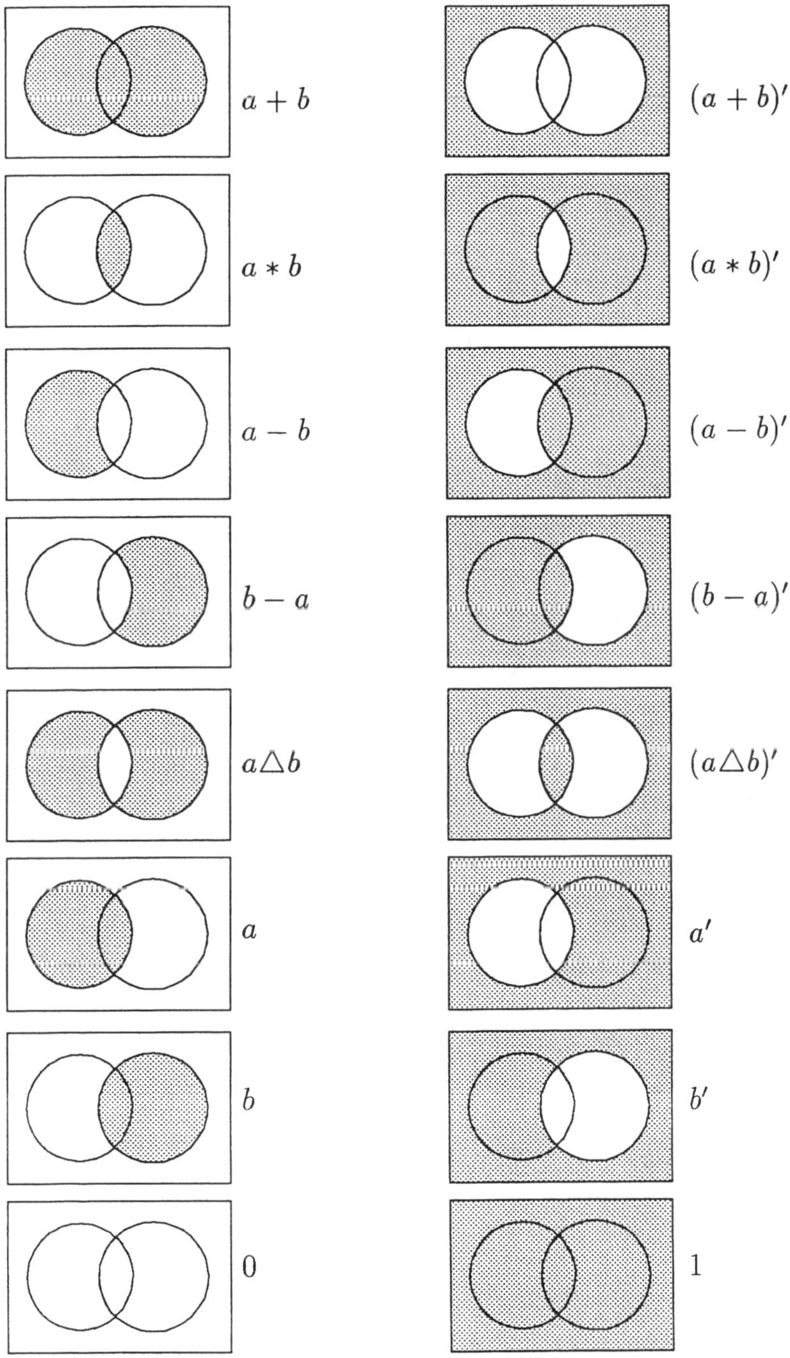

Table 5.2. Euler Venn diagrams of the 16 binary Boolean operations.

The set of all pairs $(a, b) \in B \times B$ for which $a * b = a$ (or $a + b = b$) defines a relation $(\leq)$ in B. If an element a is related to the element b we write $a \leq b$. This relation $(\leq)$ is a *partial order relation* of B, i.e. it holds that

Reflexivity :	$a \leq a$
Transitivity :	*if* $a \leq b$ *and* $b \leq c$ *then* $a \leq c$
Antisymmetry :	*if* $a \leq b$ *and* $b \leq a$ *then* $a = b$.

Rules which hold in the Boolean set algebra are of special interest. We will only enumerate them here, without proving them.

Commutativity

1a) $a + b = b + a$ 1b) $a * b = b * a$

Associativity

2a) $a + (b + c) = (a + b) + c$ 2b) $a * (b * c) = (a * b) * c$

Distributivity

3a) $a + (b * c) = (a + b) * (a + c)$ 3b) $a * (b + c) = (a * b) + (a * c)$

Idempotence

4a) $a + a = a$ 4b) $a * a = a$

Absorption

5a) $a + (a * b) = a$ 5b) $a * (a + b) = a$

Complement

6a) $a + a' = 1$ 6b) $a * a' = 0$

Doubled complement

7) $a'' = a$

De Morganean rule

8a) $(a + b)' = a' * b'$ 8b) $(a * b)' = a' + b'$

Operation with units

9a) $a + 1 = 1$ 9b) $a * 0 = 0$

10a) $a + 0 = a$ 10b) $a * 1 = a$

11a) $1' = 0$ 11b) $0' = 1$

Order relation

12) $a \leq b$ *and* $a \leq c$ *then* $a \leq b * c$

13) $a \leq b$ *then* $a \leq b + c$ *for all* $c \in B$

It is worth noticing that the axioms of a Boolean algebra are symmetric respectively to the operands $(+)$ and $(*)$. In exchanging the operations and the unit elements in a statement, the new statement is also valid. This is called the *principle of duality* . One statement is *dual* to the other. Using the order relation $(\leq)$ this principle exchanges the left and right hand side of this relation.

In our description of the modeling algebra's operations, the symmetric difference $(\triangle)$ becomes an important rule. Therefore it is necessary to consider several properties of the symmetric difference.

Lemma 5.6.
14) Commutativity: $a\triangle b = b\triangle a,$
15) Associativity: $(a\triangle b)\triangle c = a\triangle(b\triangle c),$
16) Distributivity with multiplication:

$$\begin{aligned}(a\triangle b) * c &= (a * c)\triangle(b * c)\\ (a\triangle b) * (c\triangle d) &= (a * c)\triangle(b * c)\triangle(a * b)\triangle(c * d)\end{aligned}$$

17) Negation: $(a\triangle b)' = a'\triangle b = a\triangle b',$
18) Operations with the elements 0 *and* 1*:*

$$a\triangle 0 = a, \quad a\triangle 1 = a', \quad 0\triangle a' = 1\triangle a, \quad a\triangle a' = 1 \quad a\triangle a = 0.$$

Proof. For all $a, b, c \in B$ holds

$$\begin{aligned}
a\triangle b &= (a' * b) + (a * b') = (b' * a) + (b * a') = b\triangle a,\\
(a\triangle b)\triangle c &= (((a' * b) + (a * b'))' * c) + (((a' * b) + (a * b')) * c')\\
&= ((a + b') * (a' + b) * c) + (a' * b * c') + (a * b' * c')\\
&= (a * b * c) + (b' * a' * c) + (a' * b * c') + (a * b' * c')\\
&= (a' * ((b' * c) + (b * c'))) + (a * ((b' * c) + (b * c')))'\\
&= a\triangle(b\triangle c),\\
(a\triangle b) * c &= (a' * b * c) + (a * b' * c) + (c' * c * b) + (c' * c * a)\\
&= ((a * c)' * b * c) + ((b * c)' * a * c = (a * c)\triangle(b * c),\\
(a\triangle b) * (c\triangle d) &= (a * (c\triangle d))\triangle(b * (c\triangle d)) = (a * c)\triangle(b * c)\triangle(a * b)\triangle(c * d),\\
(a\triangle b)' &= ((a' * b) + (a * b'))' = (a + b') * (a' + b) = (a' * b') + (a * b)\\
&= a'\triangle b = a\triangle b',\\
a\triangle 0 &= (a' * 0) + (a * 0') = a,\\
a\triangle 1 &= (a' * 1) + (a * 1') = a',\\
0\triangle a' &= a' = 1\triangle a,\\
a\triangle a' &= (a' * a') + (a * a) = 1 \quad \text{and}\\
a\triangle a &= (a' * a) + (a * a') = 0. \quad \square
\end{aligned}$$

It is remarkable that the symmetric difference ($\triangle$) is not distributive with respect to addition. Hence, if we work with the symmetric difference then multiplication plays an important role for decomposing formulas.

The Weak Boolean Algebra

Computation with a Boolean algebra is a favorable way of modeling. But it is often not sufficient to describe the desired operation. Therefore we are looking for abstractions which contain the Boolean algebra as well as other algebras. The question is, which properties of the Boolean algebra are essential to characterize the 3D-model or which properties can be renounced, if necessary. Let us consider the Boolean axioms. The ability to shape geometric objects by means of two binary operations should not be renounced. One should not even renounce the unit elements of the binary operations and the complement. That is, Axioms B2 and B4 should remain valid.

It is eventually possible to renounce Axioms B1 and B3. If both the operations and the complement are implemented on a computer, the terms of multiple operations will be computed each time with respect to the well-defined priority rules without using the rules of commutativity, associativity or distributivity. We have only to consider that another computation sequence of a term can produce another result.

Definition 5.7. *A set B of elements with two binary operations,* $(\dot{+})$ *and* $(\dot{*})$*, is called a* **weak Boolean algebra** $(B, \dot{+}, \dot{*})$ *if and only if the following axioms hold:*

B2' *There exists two elements of B, namely 0 and 1, as the left and right unit elements of the operation* $(\dot{+})$ *and* $(\dot{*})$*, respectively, i.e.,*

$$a \dot{+} 0 = 0 \dot{+} a = a \quad \text{and} \quad a \dot{*} 1 = 1 \dot{*} a = a.$$

B4' *For all* $a \in B$ *an element* $a' \leq B$ *exists with*

$$\begin{aligned} a \dot{+} a' &= 1 & \qquad a \dot{*} a' &= 0 \\ a' \dot{+} a &= 1 \quad \text{and} & a' \dot{*} a &= 0. \end{aligned}$$

Each function of a Boolean algebra that does not contain a constant is equal to a function in disjunctive normal form. In general, the assertion of Lemma 5.5 is not valid without restriction in a weak Boolean algebra because a transformation of any of the terms into the disjunctive normal form needs commutativity, associativity and distributivity of both operations. Lemma 5.12 is followed by Lemma 5.5. Again it is not true in general. On the other hand, such terms of disjunctive normal forms can be made in a weak Boolean algebra. The sequence of binary operations and the operands, from which such a term is composed, has indeed to be taken into consideration. Lemma 5.11 holds, and its proof is possible in the weak Boolean algebra.

The same priority rules as in the Boolean algebra are used to compute any term. If two terms of operations are not separated by parenthesis or by priority rules, they are computed from the left to the right. Permutations of operands are not permitted.

A partial order relation can be defined in a weak Boolean algebra just as in a Boolean algebra. The set of all pairs $(a, b) \in B \times B$ for which $a \dot{*} b = a$ (or $a \dot{+} b = b$) defines a relation $(\leq)$ of B. If the element a is related to the element b we write $a \leq b$.

5.2. The Natural Modeling Algebra

Now we describe another way to obtain a more abstract model than with a Boolean algebra. We provide a Boolean algebra $(B, +, *)$ only as a basis to describe the model. That allows us to produce associations between the Boolean algebra and the more an abstract model, and allows us to use both models in the process of modeling.

We obtain the starting point of this method by modifying the two Boolean operations $(+)$ and $(*)$. Let $a, b \in B$ be operands of these binary operations. A third operand $q \in B$ is added which is chosen dependent on the two operands a and b. Let the type of dependence be given by a definition of a function $q = q(a, b)$ that $B \times B$ maps into B. The special choice of this function $q(a, b)$ is described later. Notice, that this dependence need not be described by means of the Boolean algebra.

The aim is now to extend the two Boolean operations by a Boolean operation of their results and q. Thus, the result belongs to the fundamental set B again. There are several possibilities to connect $a + b$ and $a * b$ with $q(a, b)$. Let us look for a simple extension, which makes algebraic sense. Such an extension is proposed in the following definitions.

Definition 5.8. *Let $(B, +, *)$ be a Boolean algebra and $q = q(a, b)$ be a function from $B \times B$ into B. Let h_+ and h_* be binary Boolean operations. Then the operations $(\dot{+})$ and $(\dot{*})$*

$$a \dot{+} b = h_+(a + b, q(a, b)),$$
$$a \dot{*} b = h_*(a * b, q(a, b))$$

are called **modifications** *of the Boolean operations $(+)$ and $(*)$.*

Symmetry of the function $q = q(a,b)$ is not supposed. We need the following notion that gives in some sense the operand sequence of the modifications.

Definition 5.9. *An operand sequence gives a binary operation* $(\circ)$*, consisting of a finite number of binary operations* $(\dot{+})$ *and* $(\dot{*})$ *of both elements* a *and* b *and their complements. The composition* $a \circ b$ *is called a* **monotone operation** *if and only if*

> *a and a', respectively, appears isolated only as a first operand, and*
>
> *b and b', respectively, appears isolated only as a second operand,*

in each of the composition subterms with the modified operations $(\dot{+})$ *and* $(\dot{*})$.

For example, the two terms

$$a \circ b = (a \dot{*} (a \dot{+} b))' \dot{*} (a' \dot{+} b) \quad \text{and} \quad b \circ a = (b \dot{*} (b \dot{+} a))' \dot{*} (b' \dot{+} a)$$

are monotone according to their definition, but $(a \circ b) \dot{+} (b \circ a)$ is not monotone. In the special case that q is symmetric, then each term is monotone according to this definition because of (B1) of the Boolean algebra.

Two terms can be compared regarding monotonity. Let $(\circ)$ and $(\underline{\circ})$ be two monotone operations, then the both terms $a \circ b$ and $a \underline{\circ} b$ have the *same direction* and the two terms $a \circ b$ and $b \underline{\circ} a$ have a *contrary direction.*

It is obvious to use the addition and multiplication of the Boolean algebra for defining the binary operation h_+ and h_*. The arising modifications can be used for the definition of an algebra. Axioms B2 and B4 should at least be valid in this algebra. Therefore some additional conditions of the function $q = q(a,b)$ have to be assumed.

Definition 5.10. *The two modifications* $(\dot{+})$ *and* $(\dot{*})$

$$\begin{aligned} a \dot{+} b &= a + b + q(a,b), \\ a \dot{*} b &= a * b * q(a,b) \end{aligned}$$

are called **natural modifications** *of the Boolean operations* $(+)$ *and* $(*)$ *if and only if for all* $a, b, z \in B \setminus \{0,1\}$ *and each binary operation* $x = x(a,b)$*,* $y = y(a,b)$

$$\begin{aligned} q(a,z) &= q(a',z) \\ q(z,b) &= q(z,b') \end{aligned}$$

$$q(x,y) \;=\; 0 \;\; \textit{if} \;\; x \notin \{a, a', 1\} \;\; \textit{or} \;\; y \notin \{b, b', 1\}$$

$$\begin{aligned} q(0,1) &= q(1,0) = 0 \\ q(0,z) &= q(z,0) = q(0,0) = 0 \\ q(1,z) &= q(z,1) = q(1,1) = 1. \end{aligned}$$

The cases of the definition take into consideration the different significance of the following assertions. Lemma 5.12 will need the case $q = 0$, and it will be a foundation to describe special normal forms of modified binary operations. The arguments 0 and 1 of q are needed for the existence of the unit elements in the modified algebra (see Lemma 5.14).

Apart from the algebraic context, $q = 0$ and $q = 1$ are especially interesting for modeling. The possible choice of $q = 0$ permits us to modify addition and multiplication opposite to that of a Boolean operation. In this case, it holds according to the definition of q, especially

$$q(a,b) = q(a',b) = q(a,b') = q(a',b').$$

Lemma 5.11. *Let $q'(a,b)$ be the inverse element of $q(a,b)$ in the Boolean algebra, then it is valid for all $a, b, z \in B \setminus \{0,1\}$ and each binary operation $x = x(a,b)$, $y = y(a,b)$*

$$\begin{aligned} q'(a,z) &= q'(a',z) \\ q'(z,b) &= q'(z,b') \end{aligned}$$

$$q'(x,y) \;=\; 1 \;\; \textit{if} \;\; x \notin \{a, a', 1\} \;\; \textit{or} \;\; y \notin \{b, b', 1\}$$

$$\begin{aligned} q'(0,1) &= q'(1,0) = 1 \\ q'(0,z) &= q'(z,0) = q'(0,0) = 1 \\ q'(1,z) &= q'(z,1) = q'(1,1) = 0. \end{aligned}$$

Proof. The element $q(a,b)$ belongs to B for all $a, b \in B$. Because of Axiom B4, there exists a well-defined inverse element $q'(a,b)$ in the Boolean algebra $(B, +, *)$. The elements 0 and 1 are inverse to each other. The assertion follows from the definition of the natural modifications and because $u = v$ is equivalent to $u' = v'$ for all $u, v \in B$. □

Lemma 5.12. *Let $x = x(a,b)$, $y = y(a,b)$ and $z = z(a,b)$ be binary operations which are given by any terms of both elements $a, b \in B$, by means*

of the operation $(\dot{+})$ *and* $(\dot{*})$ *as well as the complement, and* $D_1 = \{a, a', 1\}$ *and* $D_2 = \{b, b', 1\}$. *It holds for the natural modifications:*
If $x \notin D_1$ *or* $y \notin D_2$ *then*

$$\begin{array}{rrcll} & x \dot{*} y & = & 0 & (\textit{disjunctivity}) \\ & x \dot{+} y & = & x + y & (\textit{replaceability}) \\ & x \dot{+} y & = & y \dot{+} x & (\textit{commutativity}). \\ \textit{It holds} & (x \dot{+} y) \dot{+} z & = & x \dot{+} (y \dot{+} z) & (\textit{associativity}) \end{array}$$

if all summands of the used sums in the equation do not belong to the set D_1 *or to the set* D_2, *respectively.*

Proof. $q(x, y) = q(y, x) = 0$ follows from the definition of $q, x \in D_1$ and $y \in D_2$. Putting this equation in the definition of natural modifications, it yields

$$\begin{array}{rcl} x \dot{*} y & = & x * y * q(x, y) = 0 \qquad \text{and} \\ x \dot{+} y & = & x + y + q(x, y) = x + y = y + x = y + x + q(y, x) = y \dot{+} x. \end{array}$$

The associativity follows from $q = 0$ according to the definition of q and the natural modifications. □

If necessary, Lemma 5.12 permits in some cases to replace the operation $(\dot{+})$ by the Boolean operation $(+)$, to exchange summands, to replace products by the element 0 and to write sums without associative parentheses.

The modifications $(\dot{+})$ and $(\dot{*})$ can also be explained as binary operations of $a, b \in B$, because the extension q is chosen dependent on the operands a and b. We produce an algebra by means of these binary operations.

Definition 5.13. *The fundamental set* B *of a Boolean algebra* $(B, +, *)$ *together with the modifications* $(\dot{+})$ *and* $(\dot{*})$ *of the operations* $(+)$ *and* $(*)$ *is called the* **modeling algebra** $(B, \dot{+}, \dot{*})$. *It is also called* **natural** *if the modifications are natural ones.*

It is not expected that the modeling algebra realizes Axioms B1 through B4 of the Boolean algebra without some additional restrictions. Indeed, two axioms hold.

Lemma 5.14. *The elements* $0 \in B$ *and* $1 \in B$ *are left and right unit elements according to the natural modifications, i.e. Axiom* B2'.

Proof. It follows from Axiom B2 of the Boolean algebra and from the definition of the natural modifications for all $a \in B$ that

$$\begin{array}{lllll} a\dot{+}0 & = & a+0+q(a,0) & = & a \\ 0\dot{+}a & = & 0+a+q(0,a) & = & a \\ a\dot{*}1 & = & a*1*q(a,1) & = & a \\ 1\dot{*}a & = & 1*a*q(1,a) & = & a. \ \square \end{array}$$

Lemma 5.15. *The inverse element a' of an element a in a Boolean algebra* $(B, +, *)$ *is the left and right inverse element of the element a in the natural modeling algebra* $(B, \dot{+}, \dot{*})$*, i.e. Axiom* B4' *and*

$$\begin{array}{lllllll} a\dot{+}a' & = & 1 & & a\dot{*}a' & = & 0 \\ a'\dot{+}a & = & 1 & \textit{and} & a'\dot{*}a & = & 0. \end{array}$$

Proof. It follows from the definition of the natural modifications and Axioms B1 and B4 of the Boolean algebra that

$$\begin{array}{lllll} a\dot{+}a' & = & a+a'+q(a,a) & = & 1 \\ a'\dot{+}a & = & a'+a+q(a',a) & = & 1 \\ a\dot{*}a' & = & a*a'+q(a,a') & = & 0 \\ a'\dot{*}a & = & a'*a*q(a',a) & = & 0. \end{array}$$

Hence, Axiom B4' of the weak Boolean algebra is valid. Suppose that two different inverse elements according to the natural modifications exist. Then two different inverse elements of the Boolean algebra exist according to the equations above. It is the opposite of the well-defined inverse elements in the Boolean algebra, i.e. the inverse elements of the Boolean algebra are the well-defined left and right inverse elements with respect to the natural modifications. □

Axioms B1 and B3 are not true without loss of generality. Lemma 5.14 and 5.15 show that $(B, \dot{+}, \dot{*})$ is a weak Boolean algebra.

The order relation $(\leq)$ of the Boolean algebra implies the order relation $(\dot{\leq})$ of the weak Boolean algebra: $a \leq b$ is equivalent to $a * b' = 0$ and $a * b' * q = 0$ according to the definition, and it follows that $a\dot{*}b = 0$ and $a\dot{\leq}b$. That is, if a pair (a, b) is ordered with respect to the Boolean algebra, then it is ordered in the natural modeling algebra.

The other direction is not true without loss of generality. But, if we define the order relation based on the equivalent equation $a * b = a$, analogously to the Boolean sense, then the so defined order relation in the natural modeling algebra implies the Boolean one. That is, the two equivalent equations $a * b' = 0$ and $a * b = a$ yield different order relations in the natural modeling algebra. We need only the first one. We can easily understand the following notions if we write a Boolean order relation, then we can write a modified order relation too.

The unit elements and the complement can be transferred (Lemmas 5.14 and 5.15). Further details of the Boolean algebra can be transferred without new assumptions of $q(a,b)$.

Lemma 5.16.
Weak idempotence:

4a)' $a\dot{+}a \geq a$ *4b)'* $a\dot{*}a \leq a$

Weak absorption:

5a)' $a\dot{+}(a\dot{*}b) \geq a$, $(a\dot{*}b)\dot{+}a \geq a$ *5b)'* $a\dot{*}(a\dot{+}b) \leq a$, $(a\dot{+}b)\dot{*}a \leq a$

Doubled complement:

7) $a'' = a.$

Proof. 4a)' holds because of

$$a\dot{+}a = a + a + q(a,a) \geq a + a = a.$$

5a)' follows from

$$a\dot{+}(a\dot{*}b) = a + (a\dot{*}b) + q(a,(a\dot{*}b)) \geq a,$$
$$(a\dot{*}b)\dot{+}a = (a\dot{*}b) + a + q((a\dot{*}b),a) \geq a.$$

4b)' and 5b)' can be obtained by means of the principle of duality. It holds because of Lemma 5.14 and 5.15:

$$\begin{array}{rclcl} a'\dot{+}(a')' &=& a'\dot{+}a &=& 1 \\ a'\dot{*}(a')' &=& a'\dot{*}a &=& 0 \end{array} \qquad \begin{array}{rclcl} (a')'\dot{+}a' &=& a\dot{+}a' &=& 1 \\ (a')'\dot{*}a' &=& a\dot{*}a' &=& 0. \end{array}$$

That is that a'' and a are left and right inverses to a'. The inverse elements are well-defined (Lemma 5.15). Hence $a'' = a$. □

Boolean rules, which are not valid without loss of generality, hold again using special assumptions.

Lemma 5.17.
Conditioned commutativity:

If $q(a,b) = q(b,a)$ *for all* $a,b \in B$ *then*

1a) $a\dot{+}b = b\dot{+}a$ *1b)* $a\dot{*}b = b\dot{*}a$.

Conditioned idempotence:

4a) if $q(a,a) = 0$ *for all* $a \in B$ *then* $a\dot{+}a = a$

4b) if $q(a,a) = 1$ *for all* $a \in B$ *then* $a\dot{*}a = a$.

Proof. Let $a,b \in B$. Because $q(a,b) = q(b,a)$,

$$a\dot{+}b = a + b + q(a,b) = b + a + q(b,a) = b\dot{+}a,$$

i.e. 1a). It follows from $q(a,a) = 0$ that

$$a \dot{+} a = a + a + q(a,a) = a,$$

i.e. 4a). 1b) and 4b) can be taken from the principle of duality again. □

Now we are interested in relations between a natural modeling algebra $(B, \dot{+}, \dot{*})$ and its Boolean algebra $(B, +, *)$. For simplification we consider, in the following context only, monotone operations of the same direction. Fundamental properties of the natural modeling algebra will appear if the natural modifications $(\dot{+})$ and $(\dot{*})$ are represented as Boolean functions with the three variables a, b and q. We consider the complete disjunctive normal form in three variables to preserve a survey of producible monotone operations of a same direction in $(B, \dot{+}, \dot{*})$. This Boolean function has the summands

$$\begin{array}{cccc} abq, & a'bq, & ab'q, & a'b'q, \\ abq', & a'bq', & ab'q', & a'b'q', \end{array}$$

describing it in the Boolean algebra $(B, +, *)$. The function $q = q(a,b)$ is the same for all summands.

Each summand of the considered complete disjunctive normal form, in the three variables a, b and q, can even be represented by a binary operation in the natural modeling algebra $(B, \dot{+}, \dot{*})$. It holds, according to the definition of the natural modifications and the de Morganean rule of the Boolean algebra, that

$$\begin{array}{rcccl} a \dot{*} b & = & abq & & \\ a' \dot{*} b & = & a'bq & & \\ a \dot{*} b' & = & ab'q & & \\ a' \dot{*} b' & = & a'b'q & & \\ (a' \dot{+} b')' & = & (a' + b' + q)' & = & abq' \\ (a \dot{+} b')' & = & (a + b' + q)' & = & a'bq' \\ (a' \dot{+} b)' & = & (a' + b + q)' & = & ab'q' \\ (a \dot{+} b)' & = & (a + b + q)' & = & a'b'q'. \end{array}$$

Euler Venn diagrams are drawn in Scheme 5.3. It holds in $(B, +, *)$ and in $(B, \dot{+}, \dot{*})$, because of Lemma 5.11 and Lemma 5.12, that the summands are disjunctive with respect to $(B, +, *)$ as well as to $(B, \dot{+}, \dot{*})$. Other binary operations in $(B, \dot{+}, \dot{*})$ can be generated by these summands by means of any finite terms of these summands.

The Boolean Axioms B1 and B3 (commutativity and distributivity) need not be valid in a natural modeling algebra. In general, terms cannot be reduced canonically in $(B, \dot{+}, \dot{*})$. Then a transformation of any term into disjunctive normal form does not exist. Now we define a new notion, particularly a notion of monotone operations, analogously to the disjunctive

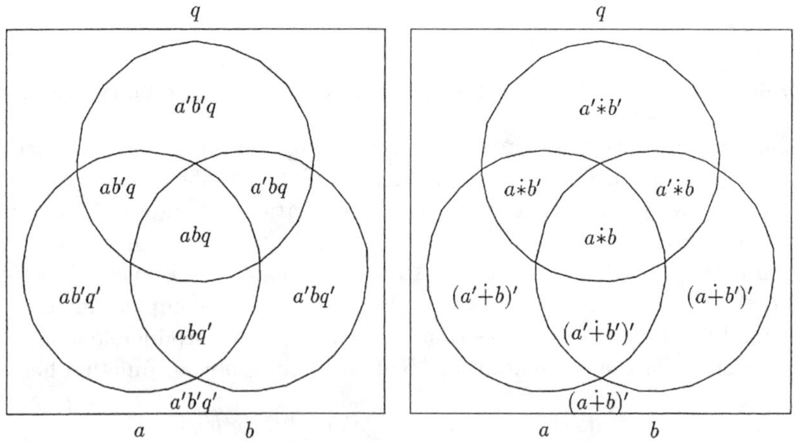

Scheme 5.3. Euler Venn diagrams of the natural modifications.

normal form of a Boolean algebra. This notion will permit us to reduce all terms of monotone operations canonically in a new way.

Definition 5.18. *A monotone operation $a \circ b$ of the natural modeling algebra $(B, \dot{+}, \dot{*})$ is given in* **weak disjunctive normal form**, *if its term is defined as a modified sum $(\dot{+})$ of terms of the form*

$$\begin{array}{cccc} a\dot{*}b, & a'\dot{*}b, & a\dot{*}b', & a'\dot{*}b', \\ (a'\dot{+}b')', & (a\dot{+}b')', & (a'\dot{+}b)', & (a\dot{+}b)', \end{array}$$

and if no two terms are identical in the modified sum.

The associativity and commutativity of such a sum of the given terms need not be taken into consideration (see Lemma 5.12). Hence, we do not pay attention to the sequence of summands and write the corresponding sums without parenthesis.

Monotone operations of the form $a \circ b$ in the natural modeling algebra can be transferred into a weak disjunctive normal form of the natural modeling algebra. For example, we reduce the term $(a \circ b)$ to a weak disjunctive normal form:

$$(a \circ b) = ((a\dot{*}b')\dot{+}(a'\dot{*}b))\dot{+}a.$$

Because of $(a\dot{*}b') \in D$ and $(a\dot{*}b') + (a'\dot{*}b) \in D$ it follows from Lemma 5.12 (replaceability) that

$$(a \circ b) = (a\dot{*}b') + (a'\dot{*}b) + a.$$

Scheme 5.3 means that $a = (a\dot{*}b') + (a\dot{*}b) + (a'\dot{+}b')' + (a'\dot{+}b)'$ and

$$(a \circ b) = (a\dot{*}b) + (a'\dot{*}b) + (a\dot{*}b') + (a'\dot{+}b')' + (a'\dot{+}b)'.$$

We get the weak disjunctive normal form of the given term

$$(a \circ b) = (a\dot{*}b)\dot{+}(a'\dot{*}b)\dot{+}(a\dot{*}b')\dot{+}(a'\dot{+}b')'\dot{+}(a'\dot{+}b)'$$

by replacing the corresponding sums regarding Lemma 5.12.

It is favorable to know the weak disjunctive normal forms of the operand a and b, the extension q and their complements.

Lemma 5.19. *The elements a, b and q have the normal forms*

$$\begin{aligned} a &= (a\dot{*}b')\dot{+}(a\dot{*}b)\dot{+}(a'\dot{+}b)'\dot{+}(a'\dot{+}b')' \\ b &= (a' * b)\dot{+}(a\dot{*}b)\dot{+}(a\dot{+}b')'\dot{+}(a'\dot{+}b')' \\ q(a,b) &= (a\dot{*}b)\dot{+}(a'\dot{*}b')\dot{+}(a'\dot{*}b)\dot{+}(a\dot{*}b'). \end{aligned}$$

The weak disjunctive normal forms of a', b', q' arise from the forms of a, b, q by changing the symbols $\dot{*}$ and $\dot{+}$ within the parentheses and by negation of the parentheses.

Proof. It follows from the definition of the natural modification, Lemma 5.14, and the laws of Boolean algebra that

$$\begin{aligned} & (a\dot{*}b')\dot{+}(a\dot{*}b)\dot{+}(a\dot{+}b)'\dot{+}(a'\dot{+}b')' \\ = \; & (a * b' * q)\dot{+}(a * b * q)\dot{+}(a' * b * q)\dot{+}(a' * b' * q) \\ = \; & (a * b' * q) + (a * b * q) + (a' * b * q) + (a' * b' * q) \\ = \; & a. \end{aligned}$$

The other forms can be reached analogously.□

The set $D = \{a, b, q, a', b', q'\}$ plays an important role in the definition of the natural modification. It is interesting to know the binary operations with operands from D. The following Lemma shows some of those operations in weak disjunctive normal form. They can be proved in the same manner as the equations in Lemma 5.19.

Lemma 5.20. *It holds with reference to the set D that*

$$\begin{aligned}
a\dot{+}q &= a\dot{+}(a'\dot{*}b)\dot{+}(a'\dot{*}b') \\
a'\dot{+}q &= a'\dot{+}(a\dot{*}b')\dot{+}(a\dot{*}b) \\
a\dot{+}q' &= a\dot{+}(a\dot{+}b')'\dot{+}(a\dot{+}b)' \\
a'\dot{+}q' &= a'\dot{+}(a'\dot{+}b)'\dot{+}(a'\dot{+}b')'
\end{aligned}$$

$$\begin{aligned}
a\dot{*}q &= (a\dot{*}b')\dot{+}(a\dot{*}b) \\
a'\dot{*}q &= (a'\dot{*}b)\dot{+}(a'\dot{*}b') \\
a\dot{*}q' &= (a'\dot{+}b)\dot{+}(a'\dot{+}b')' \\
a'\dot{*}q' &= (a\dot{+}b)\dot{+}(a\dot{+}b)'.
\end{aligned}$$

The weak disjunctive normal form of the operation $q\dot{+}b$, $q\dot{+}a$ and so on can be generated by exchanging the significance of a and b, with respect to monotonicity and Scheme 5.3. The number of possible monotone operations is very high. It is given by the possible number of weak disjunctive normal forms.

Lemma 5.21. *Each monotone operation $a \circ b$ of the natural modeling algebra $(B, \dot{+}, \dot{*})$, which does not contain a constant, is equal to a binary operation in weak disjunctive normal form of the natural modeling algebra $(B, \dot{+}, \dot{*})$. There exists* 256 *possible monotone operations in $(B, \dot{+}, \dot{*})$ of the form $a \circ b$ and* 256 *of the form $b \circ a$.*

Proof. A term of any given monotone operation of the operands a and b can be written as a Boolean term of the operands a,b and $q(a, b)$ according to the definition of the natural modifications $(\dot{+})$ and $(\dot{*})$. The new term can be reduced to a disjunctive normal form in three variables. Exchanging its summands by terms of $(B, \dot{+}, \dot{*})$, analogously to Scheme 5.3, the weak disjunctive normal form arises. The number of monotone operations $a \circ b$ in $(B, \dot{+}, \dot{*})$ results from the number of possible Boolean functions in three variables. The eight disjunctive summands can occur in the sum or not. We get $2^8 = 256$ monotone operations $a \circ b$. The extension $q(a, b)$ need not be commutative. Analogously, we get 256 monotone operations $b \circ a$.□

Assertions and schemes of binary operations without monotonicity can be deduced analogously. In that case, the complete disjunctive normal form in the four variables a, b, $q(a, b)$ and $q(b, a)$ has to be considered. The number of disjunctive summands in the complete disjunctive normal form is equal to 16, and it is possible to define 2^{16} binary operations in the natural modeling algebra.

5.3. Relations Between Boolean and Modified Operations

Similarities between a Boolean algebra and a natural modeling algebra of one and the same fundamental set B, appear if the Boolean operations $(+)$ and $(*)$ are represented in the natural modeling algebra $(B, \dot{+}, \dot{*})$ by the modified operations $(\dot{+})$ and $(\dot{*})$. It holds in the Boolean algebra that

$$\begin{aligned} a+b &= abq + a'bq + ab'q + abq' + a'bq' + ab'q' \\ a*b &= abq + abq'. \end{aligned}$$

Then we get the representation of the Boolean operations $(+)$ and $(*)$ by the modified operations $(\dot{+})$ and $(\dot{*})$ in the weak disjunctive normal form

$$\begin{aligned} a+b &= (a\dot{*}b)\dot{+}(a'\dot{*}b)\dot{+}(a\dot{*}b')\dot{+}(a'\dot{+}b')'\dot{+}(a\dot{+}b')'\dot{+}(a'\dot{+}b)' \\ a*b &= (a\dot{*}b)\dot{+}(a'\dot{+}b')' \end{aligned}$$

by means of Scheme 5.3 and Lemma 5.12.

It is worth noticing that the natural modeling algebra as well as its foundational Boolean algebra have the same fundamental set B, and the corresponding complements of each element of the fundamental set are equal to each other. The representation of the Boolean operations permits us to use the assigned Boolean rules in the natural modeling algebra, too. In this sense the Boolean algebra $(B, +, *)$ is a partial algebra of the natural modelling algebra.

Moreover, each Boolean algebra $(B, +, *)$ can be represented as a natural modeling algebra $(B, \dot{+}, \dot{*})$ by demanding

$$q(a,b) = 0$$

for all $a, b \in B \setminus \{0, 1\}$. Then we get

$$\begin{aligned} a\dot{+}b &= a+b & \qquad & & a\dot{*}b &= 0 \\ a+b &= a\dot{+}b & \qquad & \textit{and} & a*b &= (a'\dot{+}b')'. \end{aligned}$$

according to the definition of the natural modifications and the representation of the Boolean operations by modified operations. That is that addition $(\dot{+})$ will be the Boolean addition; multiplication $(\dot{*})$ disappears; and since $q(a,b) = 0$ and

$$(a'\dot{+}b')' = (a' + b')',$$

the representation of the Boolean multiplication

$$a * b = (a' + b')'$$

appears in the form of the de Morganean rule of Boolean algebra.

It is still necessary to ask for the relationship between the Boolean operations $(+)$ and $(*)$, and the modified ones $(\dot{+})$ and $(\dot{*})$. By means of a special property of the expanding operand q, we can reach distribution rules.

Definition 5.22. *The extension q of the natural modeling algebra is called* **additive** *and* **multiplicative** *if for all $a, b, c \in B$ it holds that*

$$\begin{aligned} q(a+b,c) &= q(a,c) + a(b,c) \\ q(a,b+c) &= q(a,b) + a(a,c) \\ q(a*b,c) &= q(a,c) * a(b,c) \\ q(a,b*c) &= q(a,b) * a(a,c). \end{aligned}$$

Lemma 5.23. *If the extension q is additive and multiplicative then the following distributions are valid:*

$$\begin{aligned} (a+b)\dot{+}c &= (a\dot{+}c) + (b\dot{+}c) \\ a\dot{+}(b+c) &= (a\dot{+}b) + (a\dot{+}c) \\ (a*b)\dot{*}c &= (a\dot{*}c) * (b\dot{*}c) \\ a\dot{*}(b*c) &= (a\dot{*}b) * (a\dot{*}c). \end{aligned}$$

Proof. It follows from the additivity and multiplicativity of q, and the associativity of the Boolean sum and product that

$$\begin{aligned} (a+b)\dot{+}c &= a+b+c+q(a+b,c) \\ &= (a+b)+(b+c)+q(a,c)+q(b,c) \\ &= (a\dot{+}b)+(b\dot{+}c) \\ (a*b)\dot{*}c &= a*b*c*q(a*b,c) \\ &= (a*c)*(b*c)*q(a,c)*q(b,c) \\ &= (a\dot{*}c)*(b\dot{*}c). \end{aligned}$$

The other two distributive rules follow analogously. □

5.4. The Direct and the Small Modeling Algebra

The number of possible monotone operations in a natural modeling algebra is very large. It can be restricted by special premises of $q(a,b)$. The extension q can be expressed by a modified sum over the terms

$$a\dot{*}b, \qquad a'\dot{*}b \qquad a\dot{*}b' \qquad a'\dot{*}b'.$$

Defining special subalgebras, we can assume that some of the summands of q are vanishing. It is possible to define 16 different assumptions. One of them is described as follows.

The Direct Modeling Algebra

Definition 5.24. *The extension $q(a,b)$ is called a* **direct extension** *if and only if*

$$a\dot{*}b' = a\dot{*}b = a'\dot{*}b = 0$$

for all $\quad a, b \in B \setminus \{0,1\}$.

The specification of the extension $q(a,b)$ of a binary operation demands

$$a\dot{*}b = 0 \qquad \textit{and} \qquad q(a,b) = a'\dot{*}b'$$

for all $a, b \in B$. This part of the multiplication result can be expressed as a trivial case. Defining the so arising algebra, we delete multiplication in the natural modeling algebra $(B, \dot{+}, \dot{*})$ by exchanging its result for zero and take another non-trivial operation to explain the new algebra in an analogous form as the natural modeling algebra.This definition contains the non-zero result

$$a'\dot{*}b' = q(a,b)$$

as the non-vanishing part of the multiplication in the natural modeling algebra.

Definition 5.25. *Let* $(B, \dot{+}, \dot{*})$ *be a natural modeling algebra with a direct extension* $q(a,b) \in B$ *for all* $a, b \in B$. *We define the algebra* $(B, \acute{+}, \acute{*})$ *with the fundamental set* B *and the operations*

$$\begin{aligned} a\acute{+}b &= a\dot{+}b \\ a\acute{*}b &= a'\dot{*}b'. \end{aligned}$$

Then the algebra $(B, \dot{+}, \dot{*})$ *and* $(B, \acute{+}, \acute{*})$, *respectively, is called a* **direct modeling algebra**.

Clearly, the direct modeling algebra is a weak Boolean algebra because it is a natural modeling algebra, according to the operation $(\dot{+})$ and $(\dot{*})$ and the unit elements of the Boolean algebra. With respect to the operation $(\acute{+})$ and $(\acute{*})$, the rules of a weak Boolean algebra are valid without the existence of a unit element of the exchanged multiplication $(\acute{*})$. According to Scheme 5.3, we get parts of disjunctive normal forms of terms in the direct modeling algebra.

Definition 5.26. *A monotone operation $a \circ b$ is given in* **direct disjunctive normal form** *if its term is represented as a sum $(\acute{+})$ over the five terms*

$$(a'\acute{+}b')', \quad (a\acute{+}b')', \quad (a'\acute{+}b)', \quad (a\acute{+}b)' \quad \text{and} \quad (a\acute{*}b),$$

and if no two terms are identical in the sum.

All monotone operations of the direct modeling algebra can easily be demonstrated. Tables 5.5 and 5.6 show their Euler Venn diagrams. The element $q \in B$ is not drawn as a connected area (see Scheme 5.4) in order to clarify the method presented in the following chapters.

Table 5.5 shows the first half of all possible Euler Venn diagrams. They correspond to the diagrams of Table 5.2 of the Boolean algebra $(B, +, *)$. The Boolean algebra is a partial algebra of the direct modeling algebra, since

$$\begin{aligned} a + b &= (a'\dot{+}b')'\dot{+}(a\dot{+}b')'\dot{+}(a'\dot{+}b)' \\ a * b &= (a'\dot{+}b')', \end{aligned}$$

and therefore

$$\begin{aligned} a + b &= (a'\acute{+}b')'\acute{+}(a\acute{+}b')'\acute{+}(a'\acute{+}b')' \\ a * b &= (a'\acute{+}b')'. \end{aligned}$$

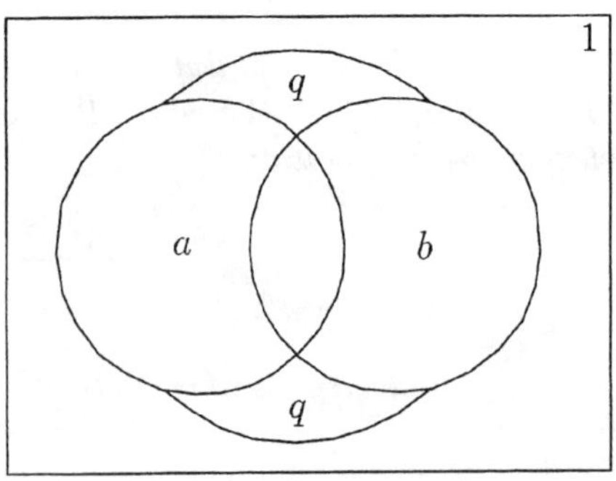

Scheme 5.4. Symbolic representation of the sets a, b, q and the unit element 1 for demonstrating the monotone operations of same direction in the direct modeling algebra.

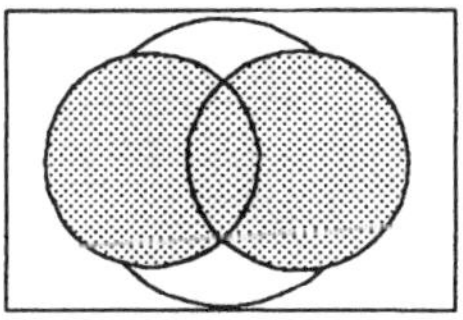	$a + b$	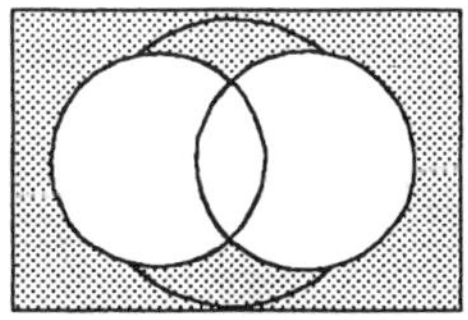	$(a + b)'$
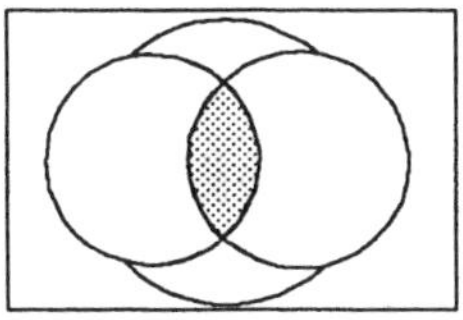	$a * b$	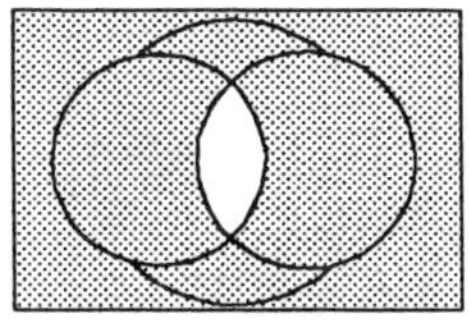	$(a * b)'$
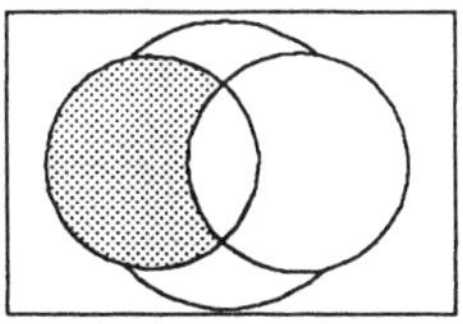	$a - b$	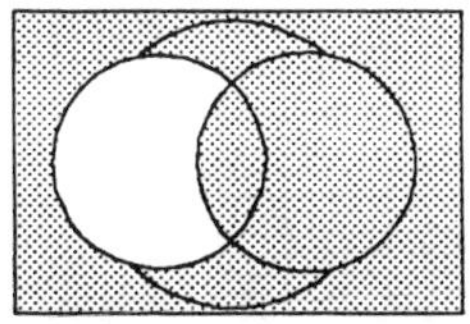	$(a - b)'$
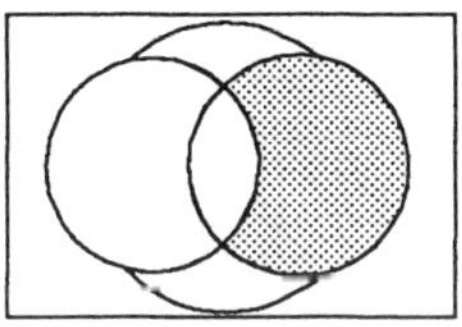	$b - a$	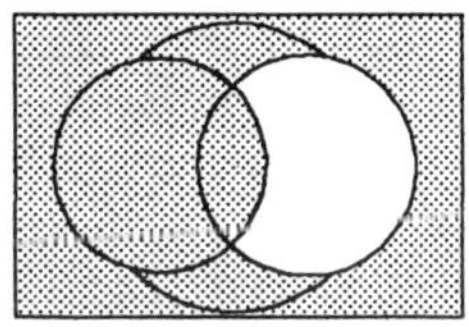	$(b - a)'$
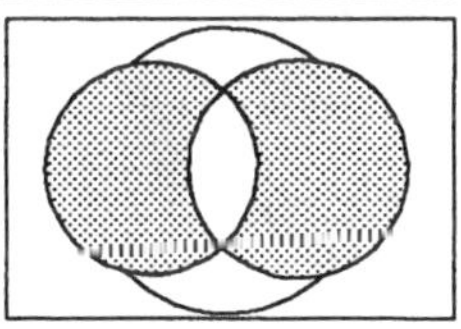	$a \triangle b$	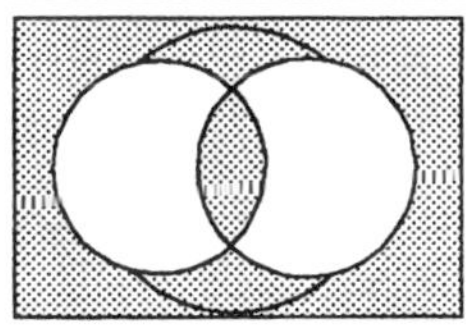	$(a \triangle b)'$
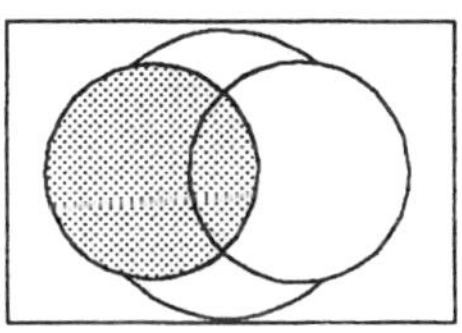	a	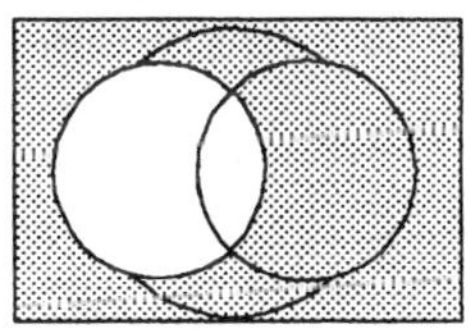	a'
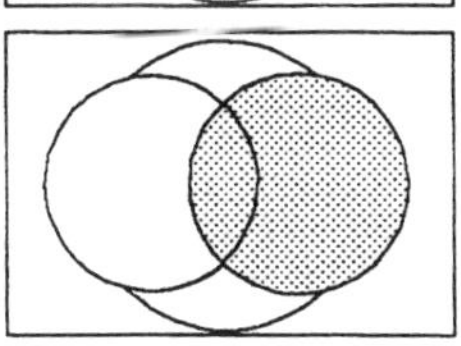	b	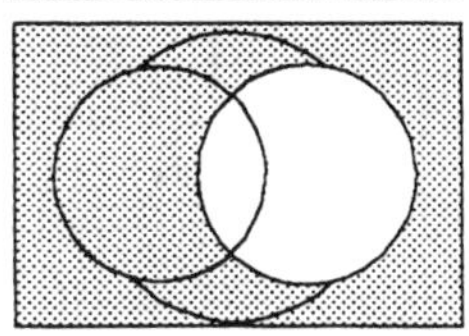	b'
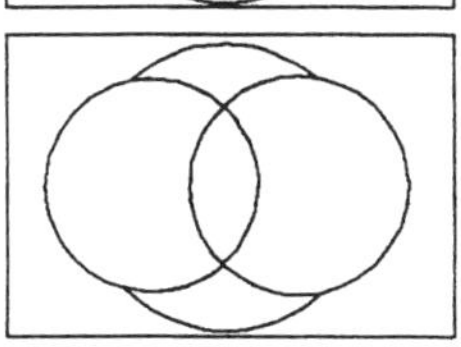	0	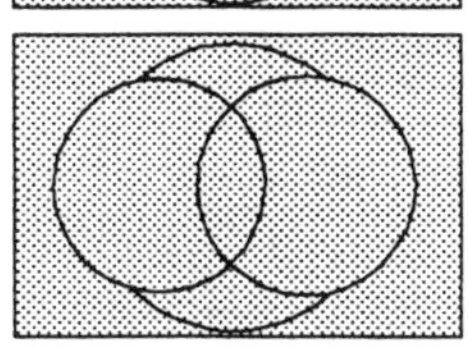	1

Table 5.5. Diagrams from the first half of monotone operations with the same direction in a direct modeling algebra.

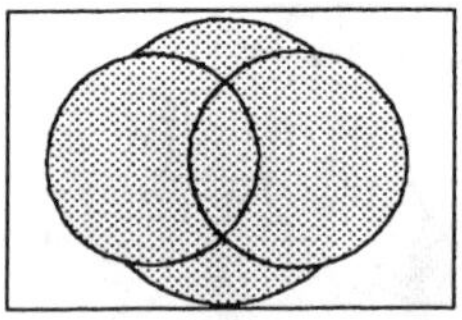	$a \acute{+} b$	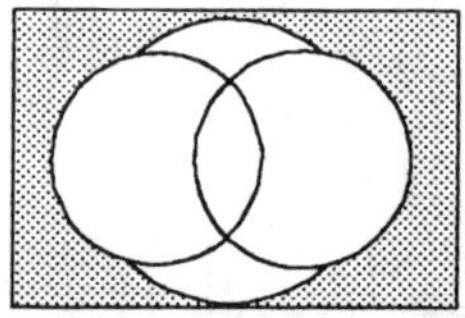	$(a \acute{+} b)'$
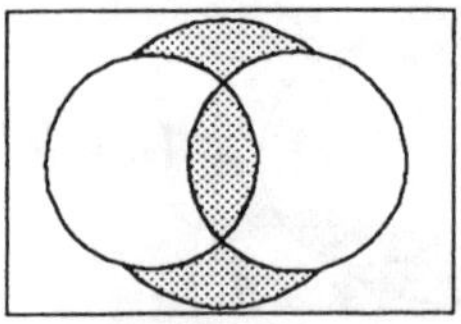	$a \acute{*} b$	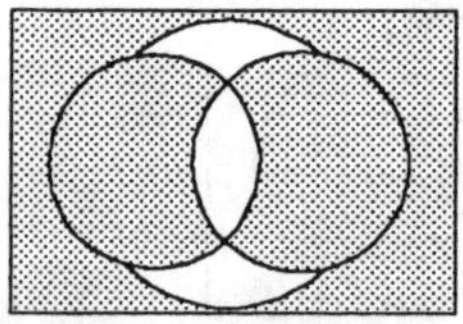	$(a \acute{*} b)'$
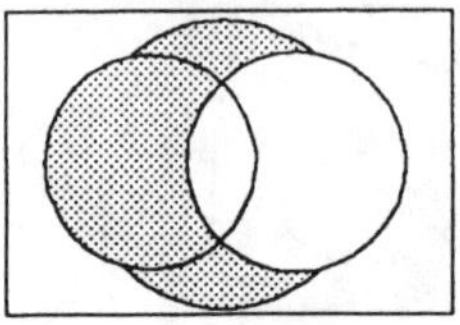	$a \acute{-} b$	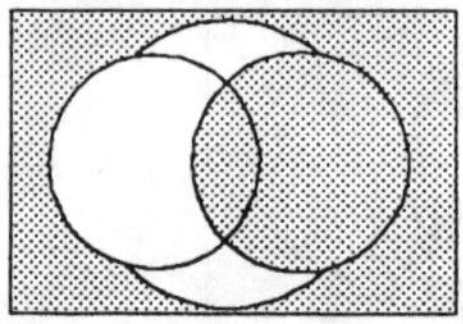	$(a \acute{-} b)'$
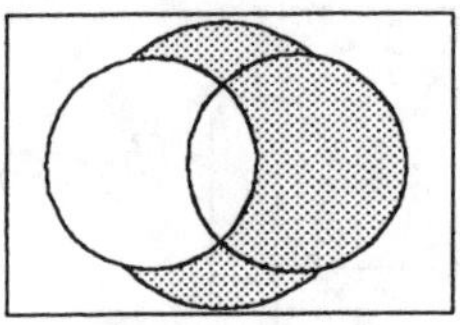	$b \acute{-} a$	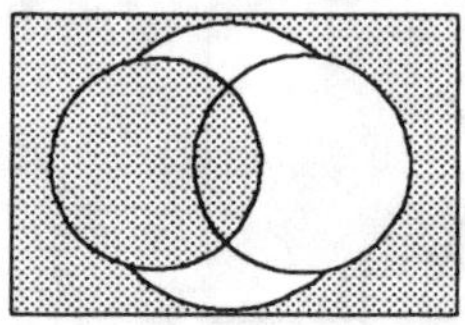	$(b \acute{-} a)'$
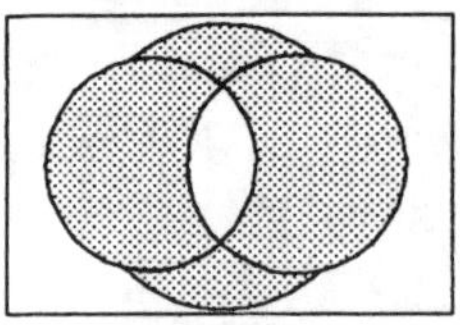	$a \acute{\triangle} b$	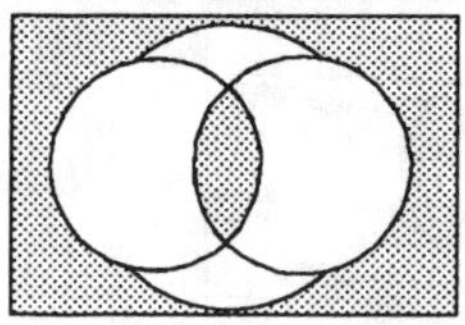	$(a \acute{\triangle} b)'$
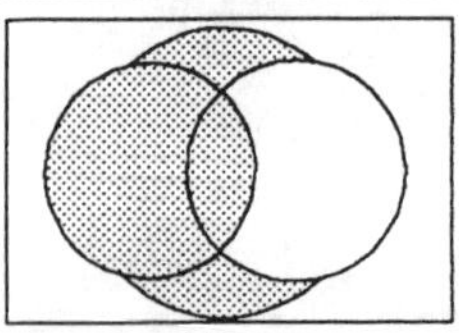	$\acute{a}$	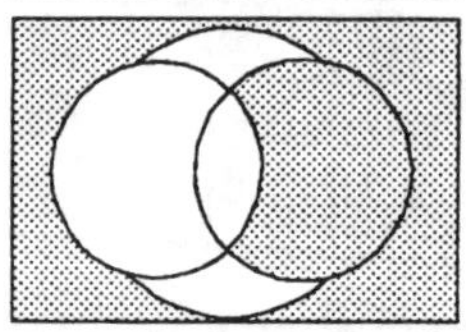	$\acute{a}'$
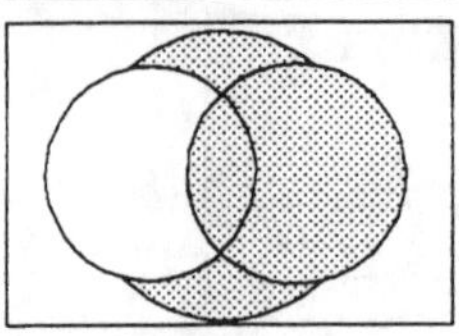	$\acute{b}$	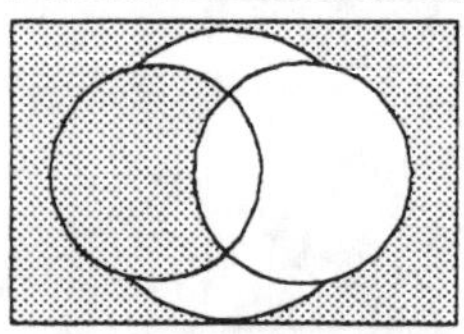	$\acute{b}'$
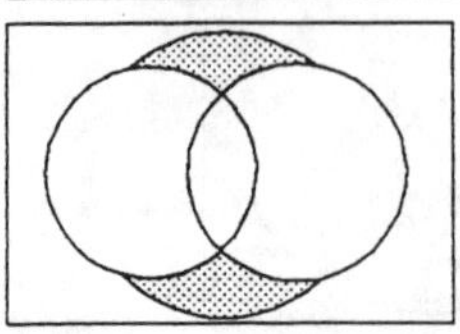	$\acute{0}$	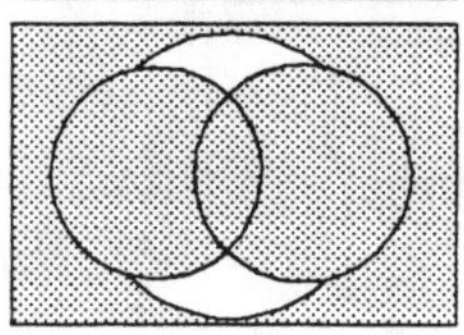	$\acute{1}$

Table 5.6. Diagrams from the second half of monotone operations with the same direction in a direct modeling algebra.

Table 5.6 shows the rest of the monotone operations with the same direction in a direct modeling algebra. The difference $a\acute{*}b'$ is characterized by $a\acute{-}b$, the difference $a'\acute{*}b$ is characterized by $(b\ \acute{}\ a)$ and we write $a\acute{\bigtriangledown}b$ instead of the symmetric difference $(a\acute{-}b)\acute{+}(b\acute{-}a)$.

The Small Modeling Algebra

Substructures of a natural modeling algebra are interesting. They can be described, for example, if the new subalgebra is defined by means of different two binary operations from the set of all 256 possible monotone operations with the same direction in the natural modeling algebra. In this sense, we exchange the multiplication $a\dot{*}b$ in the definition of the natural modeling algebra by the operation $(a'\dot{+}b')'$.

Definition 5.27. *Let $(B, \dot{+}, \dot{*})$ be a natural modeling algebra. We define the algebra $(B, \hat{+}, \hat{*})$ with the fundamental set B and the operations $(\hat{+})$ and $(\hat{*})$*

$$\begin{aligned} a\hat{+}b &= a\dot{+}b \\ a\hat{*}b &= (a'\dot{+}b')'. \end{aligned}$$

It is called a **small modeling algebra.**

In the notion of the Boolean algebra $(B, +, *)$, it holds for the two operations $(\hat{+})$ and $(\hat{*})$ that

$$\begin{aligned} a\hat{+}b &= a + b + q(a, b) \\ a\hat{*}b &= a * b * q'(a, b). \end{aligned}$$

Lemma 5.28. *The elements $0 \in B$ and $1 \in B$ are left and right unit elements according to the operations $(\hat{+})$ and $(\hat{*})$ of the small modeling algebra $(B, \hat{+}, \hat{*})$, i.e.* B2'.

Proof. It holds according to the definition of the operation $(\dot{+})$ that

$$\begin{aligned} a\hat{+}0 &= a\dot{+}0 = a \qquad \textit{and} \\ 0\hat{+}a &= 0\dot{+}a = a. \end{aligned}$$

It holds for the unit element 1 of the Boolean algebra and in conformity with Lemma 5.11 that

$$\begin{aligned} a\hat{*}1 &= (a'\dot{+}1')' = (a'\dot{+}0)' = (a')' = a \qquad \textit{and} \\ 1\hat{*}a &= (1'\dot{+}a')' = (0\dot{+}a')' = (a')' = a. \ \square \end{aligned}$$

Lemma 5.29. *The inverse element a' of the Boolean algebra $(B,+,*)$, which is inverse to $a \in B$ is the well-defined left and right inverse element of the element a in the small modeling algebra $(B,\hat{+},\hat{*})$, i.e.* B4', *and*

$$\begin{array}{rclcrcl} a\hat{+}a' &=& 1 & & a\hat{*}a' &=& 0 \\ a'\hat{+}a &=& 1 & \text{and} & a'\hat{*}a &=& 0. \end{array}$$

Proof. It follows from the definition of the operations $(\hat{+})$ and $(\hat{*})$, as well as from Lemma 5.15 that

$$\begin{aligned} a\hat{+}a' &= a\dot{+}a' = 1 = a'\dot{+}a = a'\hat{+}a \quad \text{and} \\ a'\hat{*}a &= (a\dot{+}a')' = 1' = 0 = (a'\dot{+}a)' = a\hat{*}a'. \end{aligned}$$

The inverse element in a natural modeling algebra is well-defined and it yields the well-defined inverse element in the small modeling algebra. □

Lemmas 5.14 and 5.15 show that the small modeling algebra $(B,\hat{+},\hat{*})$ is a weak Boolean algebra. Analogously, the validity of Lemmas 5.11, 5.14 and 5.15 can be transferred into the small modeling algebra. As opposed to the natural modeling algebra, we have the following Lemma:

Lemma 5.30. *The De Morganean rules are valid in the small modeling algebra:*

8a)' $(a\hat{+}b)' = a'\hat{*}b'$ and 8b)' $(a\hat{*}b) = a'\hat{+}b'$.

Proof. It follows from the definition of $(\hat{*})$ and $(\hat{+})$, as well as Lemma 5.15, that

$$a'\hat{*}b' = (a\dot{+}b)' = (a\hat{+}b)' \quad \text{and} \quad (a\hat{*}b)' = (a'\dot{+}b')'' = a'\dot{+}b'. \quad \square$$

The small modeling algebra $(B,\hat{+},\hat{*})$ is a partial algebra of the natural modeling algebra because its operations are defined as operations of the natural modeling algebra. Hence, its monotone operations of the form $a \circ b$ can be written in weak disjunctive normal form. Indeed, not every monotone operation of the natural modeling algebra can be represented in the small modeling algebra.

Because the multiplication, $(\hat{*})$ as well as the addition $(\hat{+})$, are defined only by means of the addition $(\dot{+})$ and the complement $(')$, one could ask which monotone operations can be represented only by the operation $(\dot{+})$ and the complement $(')$ in the natural modeling algebra $(B,\dot{+},*)$, and hence in the small modeling algebra too.

Lemma 5.31. *Each monotone operation $a \circ b$ of the natural modeling algebra, which can be represented by a term of addition $(\dot{+})$ and the complement, can be represented as a sum $(\dot{+})$ over the five summands*

$$(a'\dot{+}b')', \quad (a\dot{+}b')', \quad (a'\dot{+}b)', \quad (a\dot{+}b)' \quad \textit{and} \quad q.$$

Proof. Lemma 5.11 permits the selection of sums without a common complement respectively within a common complement in a weak disjunctive normal form, which only consists of sums $(\dot{+})$. One can still ask for a transformation of a common complement in a weak disjunctive normal form, which only consists of sums $(\dot{+})$. The unit element 1 can be represented as the following Boolean sum

$$1 = (a'\dot{+}b')' + (a\dot{+}b')' + (a'\dot{+}b)' + (a\dot{+}b)' + q$$

with disjunctive summands. For example, let

$$x = (a'\dot{+}b')' + (a\dot{+}b')',$$

then it yields

$$x' = (a'\dot{+}b)' + (a\dot{+}b)' + q$$

because of Lemma 5.29 and the given representation of the unit element 1. Analogously, all complements of a weak disjunctive normal form can be expressed by a Boolean sum They can be written by means of Lemma 2.11 as modified sums $(\dot{+})$ again. Together we get the validity of Lemma 5.31. □

The sums $(\dot{+})$ and $(\hat{+})$ are identically defined. Therefore, Lemma 5.31 can be used as a base for describing canonical forms of monotone operations with a same direction in the small modeling algebra. We define analogously to the weak disjunctive normal form:

Definition 5.32. *A monotone operation $a \circ b$ is given in* **small disjunctive normal form** *if its term is represented as a sum $(\hat{+})$ over the five summands*

$$(a'\hat{+}b')', \quad (a\hat{+}b')', \quad (a'\hat{+}b)', \quad (a\hat{+}b)' \quad \textit{and} \quad q,$$

and if no two terms are identically in the sum.

We get a possible 32 monotone operations with the same direction in a small modeling algebra $(B, \hat{+}, \hat{*})$ by means of Lemma 5.31. It follows from five possible summands, each of which can or cannot appear as a summand in a small disjunctive normal form.

The Tables 5.7 and 5.8 show the Euler Venn diagrams of the monotone operations with the same directions, which can be generated in the small modeling algebra. The diagrams of Table 5.8 are different from the diagrams of Table 5.7 only by the opposite sense of the occurrence of the extension q in each result. Hence an analogous notion is chosen for the operations, which is characterized in Table 5.8 by means of the symbol $(\check{\ })$.

5.5. Basic Operations

It is not necessary to work with all binary operations during the modeling process or in the implementations of the modification algebras. The expense would be to high. It is even enough to work only with a few of them, the so-called basic operations. One discovers such basic operations that can directly generate all of the other binary operations by means of the algebra axioms.

Lemma 5.33. *The operation sets*

$\{(+),(*),(')\}$ *or* $\{(+),(')\}$	*in the Boolean algebra,*
$\{(\dot{+}),(\dot{*}),(')\}$	*in the natural modeling algebra and*
	in the direct modeling algebra and
$\{(\hat{+}),(')\}$	*in the small modeling algebra*

define, in each case, a system of basic operations for all (monotone) operations.

Proof.
Boolean algebra:
The set $[(+),(*),(')]$ as a base of the Boolean algebra is given by Axiom B1 through B4. The basis $[(+),(')]$ arises from substitutions of multiplication by means of the De Morganean Rule.

Natural and direct modeling algebra:
All of the monotone operations can be written by $(\dot{+}),(\dot{*})$ and $(')$, in weak disjunctive normal form, and therefore the set of these operations is a basic system. The operation $(\dot{*})$ cannot only be described by addition and the complement because of the absence of the De Morganean Rules according to Scheme 5.3.

Small modeling algebra:
Each operation can be represented in the natural modeling algebra. The set $[(\dot{+}),(\dot{*}),(')]$ is also a basic system of the natural modeling algebra. Nevertheless, multiplication does not play a role in the definition of the

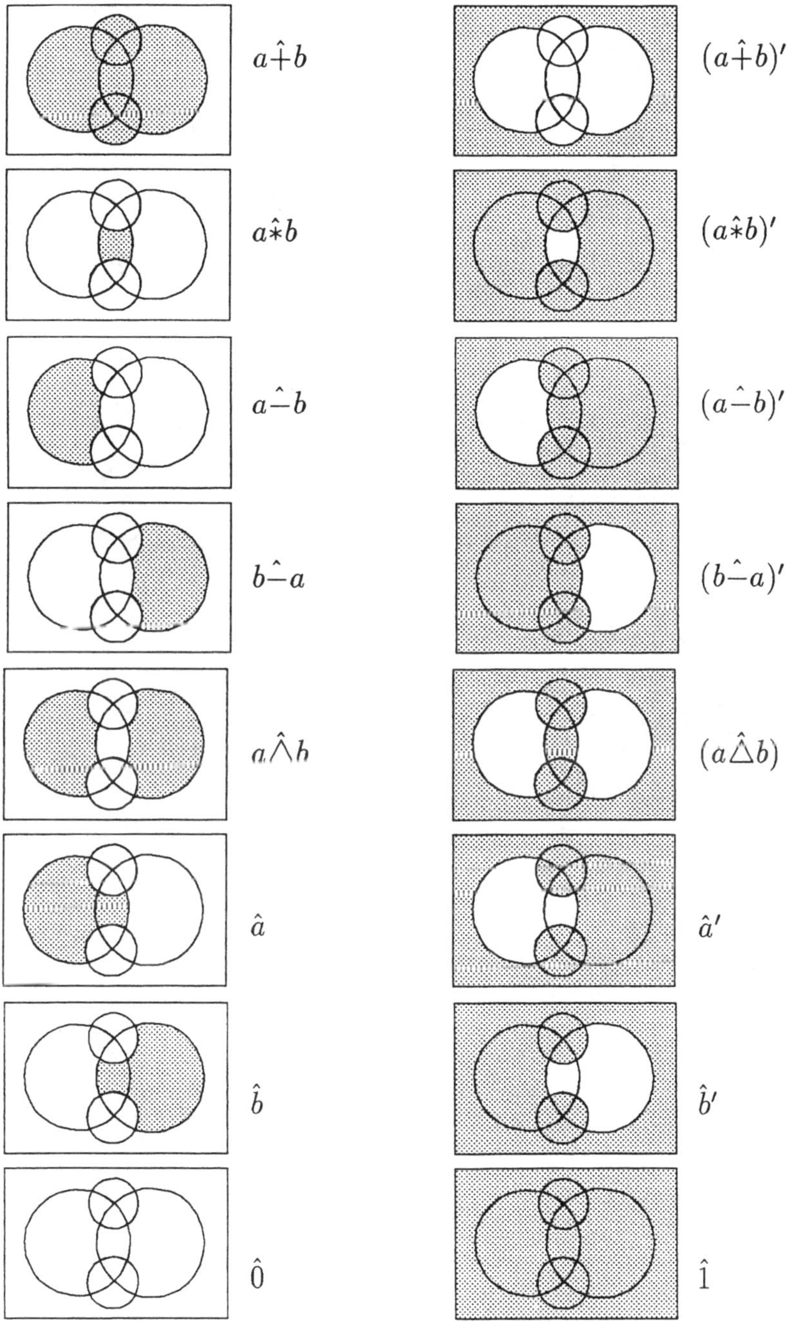

Table 5.7. Diagrams from the first half of monotone operations with the same directions in the small modeling algebra.

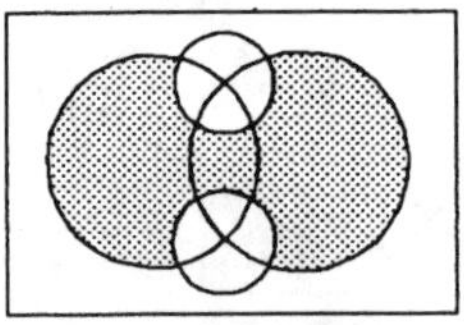	$a\check{+}b$	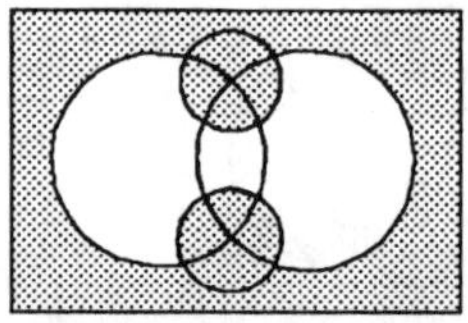	$(a\check{+}b)'$
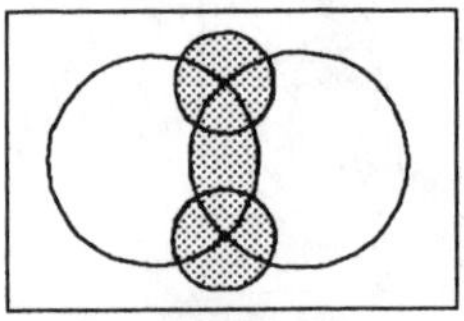	$a\check{*}b$	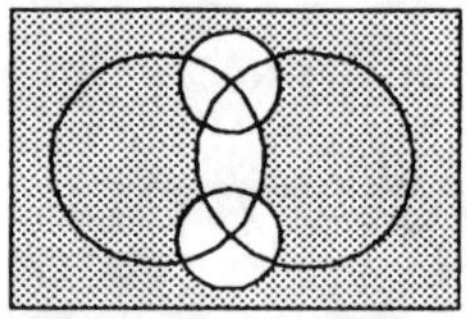	$(a\check{*}b)'$
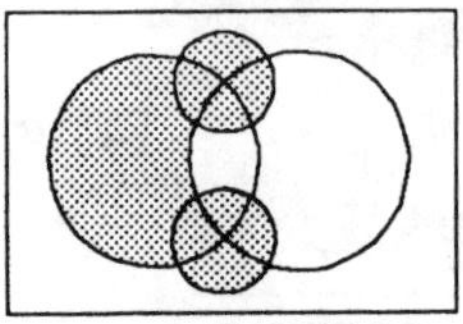	$a\check{-}b$	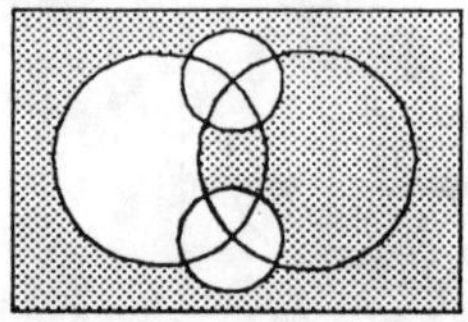	$(a\check{-}b)'$
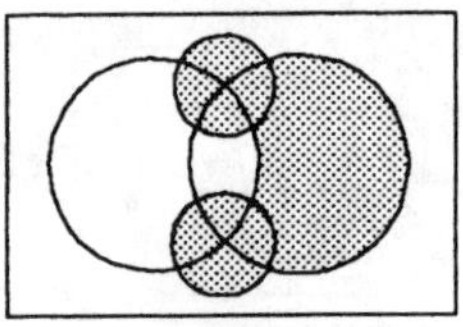	$b\check{-}a$	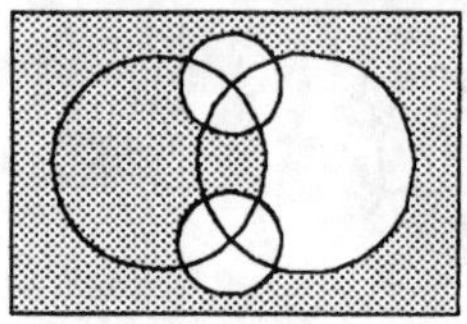	$(b\check{-}a)'$
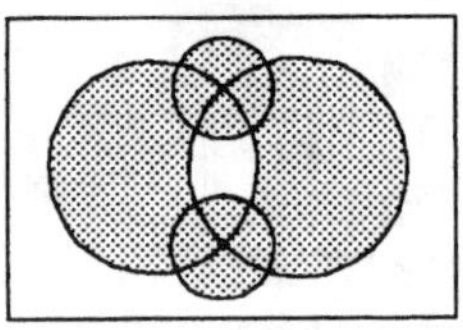	$a\check{\triangle}b$	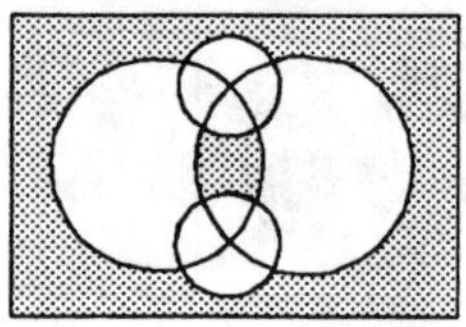	$(a\check{\triangle}b)$
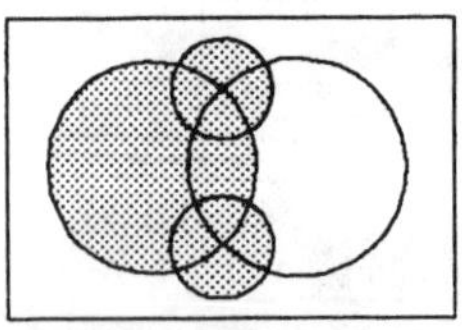	$\check{a}$	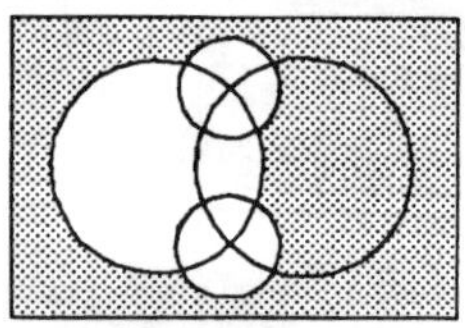	$\check{a}'$
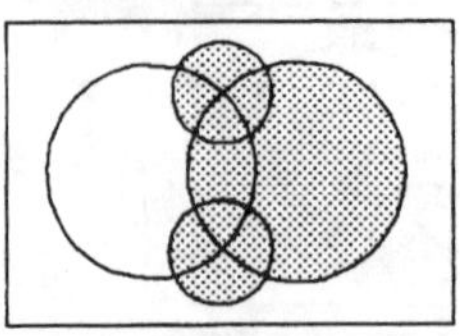	$\check{b}$	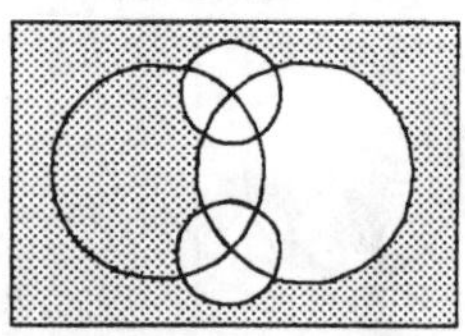	$\check{b}'$
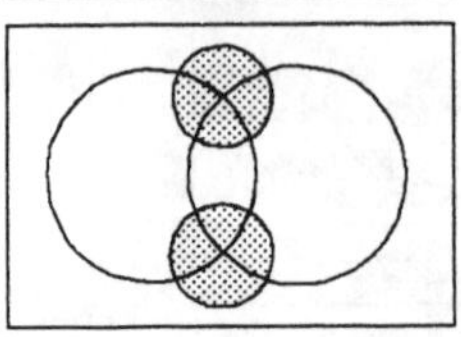	$\check{0}$	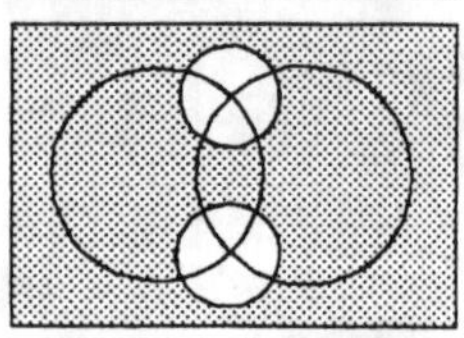	$\check{1}$

Table 5.8. Diagrams from the second half of monotone operations with the same directions in the small modeling algebra.

algebra. Both operations $(\hat{+})$ and $(\hat{*})$ are described by the operations $(\dot{+})$ and $(')$. The operation $(\dot{+})$ and $(\hat{+})$ are identical ones. The set $[(\hat{+}), (')]$ can therefore be taken as a basic system. □

Lemma 5.33 has to be regarded from two points of view. On the one hand, it gives us the possibility to most simply describe all (monotone) operations canonically by using a small number of basic elements. The user of such algebras has another view. He does not absolutely want to work at this low level. Therefore, it is sensible on the other hand to extend the basic system in a way that it is adapted to the user's thinking. In this sense, one likes to work in modeling systems with addition, multiplication, complement and difference (i.e., multiplication with complement).

Certainly, the designer of the internal operations of a system with a natural modeling algebra will use the well-known Boolean set operations and additionally the operation (q) that defines the binary operation of a and b with the extension $q = q(a, b)$ as the result. The following lemma holds because each binary operation with the operands a and b in a natural modeling algebra can be computed by a trinary Boolean one with the three operands a, b and q.

Lemma 5.34. *The operation sets* $\{(+), ('), (q)\}$ *and* $\{(*), ('), (q)\}$ *define, in each case, a system of basic operations for all (monotone) operations of a natural modeling algebra* $(B, \dot{+}, \dot{*})$.

6

Computation of Algebraic Manifolds

A computer can store only a finite amount of information, i.e., substantially less compared to what is mathematically possible. We must therefore specify the domains of computed numbers and how we have to calculate using these numbers. The following view supports fixed-point numbers for describing computer internal models. Since we want to achieve a well-defined fixed-point precision, it is often necessary to use a large number of digits.

6.1. Number Domains

Let us consider domains of numbers describing the representation of geometric objects and their calculation. Representation numbers often differ from numbers used for computing the representation. At first we have to start from the desired representation domains because they are the base of the computation.

The points, curves and surfaces of a boundary description are mathematically defined in Euclidean point space E^3. They should reflect the geometric forms of technical objects for example. Tools and products may extend for some meters and differ from one another by as little as micrometer. This yields, in millimeters, a number of the form

$$\pm XXX.XXX$$

of seven decimals. Six decimals are enough sometimes. In the following context we assume we have numbers with no fewer than six valid decimals.

It is a standard practice to add an exponent and to normalize the valid decimals, so that the decimal point can be found in front of the first valid digit (in other contexts, sometimes after the first). Such normalized numbers are floating-point numbers. Near zero, there is a relatively large interval which contains, without zero, only one positive number. The neighboring interval of the same length is very compactly filled with floating-point numbers. This density decreases in the direction of the largest floating-point number. Therefore, an approximation of E^3 by means of floating point numbers seems to be undesirable. Because of this non-regular number distribution, points nearly at the origin would reach a higher precision than points far away from the origin.

The rounding precision of the real number $x \neq 0$

$$x = (sign x)(0.d_1 ... d_m d_{m+1} ...)10^k$$

with $d_1 \geq 0$ is defined by rounding x to a *floating-point number* y

$$y = rd_m(x) = (sign x)(0.d_1 ... d_{m-1} d'_m)10^k$$

and

$$d'_m = \begin{cases} d_m & if \;\; d_{m+1} \leq 4 \\ 1 + d_m & otherwise \end{cases}$$

by m valid decimals with the *absolute error*

$$\mid rd(x) - x \mid \leq 0.5 \cdot 10^{k-m}$$

and the *relative error*

$$\mid (rd(x) - x)/x \mid \leq 0.5 \cdot 10^{-m}.$$

In particular this means that the relative error and absolute error are equal to each other if we use numbers without an exponent, i.e., $k = 0$. Then we can use floating-point arithmetic as well as fixed-point arithmetic in the same sense. The relative error is mostly expressed by the tolerance $(1 + \varepsilon)$ of the rounding procedure

$$rd(x) = (1 + \varepsilon)x \;\; and \;\; \mid \varepsilon \mid \leq 0.5 \cdot 10^{-m}$$

to represent the real number x in actual computation.

Using fixed-point numbers we have to consider the location of the decimal point. It is all the same from the viewpoint of input, because the input of numbers is associated with a conversion in any case. We define our real number by a fixed-point representation of an even number m of decimals,

$$\underset{-}{+}XXX \quad \underset{m/2}{\cdots} \quad X.X \quad \underset{m/2}{\cdots} \quad XXX$$

a sign and the decimal point in the middle. The smallest nonvanishing positive number and the largest number are

$$y_{min} = 000...0.0...001$$
$$y_{max} = 999...9.9...999$$

and we have

$$\frac{1}{y_{max}} = y_{min} + \frac{y_{min}}{y_{max}} \approx y_{min}.$$

Describing points in E^3 by such fixed-point numbers, this relation reflects the similarity of the origin and the far point which closes E^3 to E^{-3} with respect to the stereographic projection of E^3. In that sense, the overflow of the multiplication corresponds to division by zero.

In contrast to floating-point numbers, it is now not necessary to distinguish between the absolute error

$$\mid rd(x) - x \mid \leq 0.5 \cdot 10^{-m/2}$$

and the relative error

$$\mid \frac{rd(x) - x}{x} \mid \leq 0.5 \cdot 10^{-m/2}$$

We define the *representation precision* by

$$EPS = 0.5 \cdot 10^{-m/2}.$$

In the following context, the number m denotes the number of fixed decimals as well as the possible number of valid decimals in a floating-point representation.

Sometimes the rounding rule of real numbers near y_{min} and y_{max} takes the form

$$rd(x) = \begin{cases} 0 & if \quad |x| < y_{min} \\ (sign\ x) y_{max} & if \quad |x| > y_{max}. \end{cases}$$

The first case means underflow, the second one means overflow. If it is possible by software or by hardware, the rounding of underflow and overflow should be adapted to the rounding explained above. Hence, the representation precision EPS of a real number in fixed-point representation is the smallest positive number which can be added to any fixed-point number and changing its value.

Depending on a particular application, different number domains may be used for precision considerations. The different precision levels influence the modeling process. The user probably wants to enhance the software with a definable precision (calling this the representation precision EPS). Making this available for communication between user and software, calculations with higher precision are necessary (for computation precision and interpolation precision).

The so-called *number domain* of the modeling process uses fixed-point numbers with a length m as a rule. The *interpolation domains* or *computation domains* are collections of floating-point numbers with $2m$ valid decimals. A result of a computation can thus have different precision. They are called *suitable results* if their absolute error is smaller than the precision in the definition domain.

There are problems and algorithms that can be represented without rounding errors. We assign to a specific task the so called *correct number domain.* For some special cases, the computation domains are already free of rounding errors. Computations without rounding need many valid digits as a rule, if they are indeed possible. There are difficult problems which cannot be managed by increasing the number of digits, i.e., the results can be non-suitable although the algorithms are well-chosen. It depends on the specific task. A geometric task may have to be defined differently, in the sense of precision, not only by means of geometric motivations.

Discretization of Euclidean Space E^3

The mathematical model of Euclidean space E^3 cannot completely be transferred into the internal computer model because of the finite amount of representable numbers.

Definition 6.1. *Let $DE^3(m)$ be the set of all points (x, y, z) of Euclidean point space E^3 whose components x, y, z have a centered and fixed-number representation of an even length m. The set $DE^3(m)$ is called the* **discretization of E^3**. *Corresponding to higher dimensions, $DE^n(m)$ is generally called a discretization of E^n.*

We will attempt to use the advantages of floating-point arithmetic together with that of the fixed representation. The normalization, that is often required of intermediate results, can be avoided. Computations are to be described in n-dimensional space R^n. Without loss of generality, we take a number representation of the base 10. We denote respectively the smallest and the largest possible exponent of the floating-point number by $\bar{k}$ and $-\bar{k}$, assuming that the limits of the exponents are sufficient to describe the normalization of intermediate results in the set of floating-point numbers.

Definition 6.2. *The set $DR^n(m)$ contains all points of R^n whose components have a floating-point representation with length m. The set $DR^n(m)$ is called a* **discretization of $\boldsymbol{R}^n$** .

In particular, we are now looking for a suitable discretization of R^3 that is compatible to the computation with elements of $DE^3(m)$. A suitable discretization, in our sense, is one that reflects all of the differences of the accurately represented decimals, in order to distinguish between different kinds of parameters as well as to represent the desired objects in a stable way during computation. Efficient domains arise if we try to represent all of the possible computations in E^3 without number domain overflow. Possible computations in E^3 are the difference of vectors, the scalar product and the vector product.

As the first operation in E^3, we consider the computation of the scalar product as a foundational operation. We try to compute significant scalar products without rounding errors by means of doubling the length of significant digits, without a number domain overflow.

Definition 6.3. *Let be $M_1 = \{(x, y, z) \in E^3 \mid x^2 + y^2 + z^2 < 10^m\}$. The set $DE^3(m) \cap M_1$ is called a* **restricted discretization** *of E^3.*

Lemma 6.4. *If $a, b \subset DE^3(m) \cap M_1$ then their scalar product (a, b) and their vector product $[a, b]$ are computable without error in the $DR^1(2m)$ and $DR^3(2m)$, respectively.*

Proof. We compute the scalar product

$$(a, b) = a_1 b_1 + a_2 b_2 + a_3 b_3$$

and the vector product

$$[a, b] = (a_2 b_3 - a_3 b_2, a_3 b_1 - a_1 b_3, a_2 b_1 - a_1 b_2)$$

using fixed-point numbers with exact arithmetic. Then, the number of digits after the decimal point can at most be doubled. But the number of digits needed before the decimal point is not clear. It follows from the Cauchy Schwarz inequality and the restriction of $a, b \in DE^3(m) \cap M_1$

according to its definition

$$(a,b)^2 \leq (a,a)(b,b) < 10^{2m}$$

and

$$(a,b) < 10^m.$$

The number of digits before the decimal point for the scalar product can at most be doubled. The single components of the vector product can be represented as special scalar products. □

The inequality of the last definition, restricting $DE^3(m)$, carves out the interior of a sphere from E^3, see Figure 6.1. This sphere borders all boundary surfaces of $DE^3(m)$. The space is not practical. Using a cubic domain, the maximal length in the axis directions can easily be handled assuming a Cartesian coordinate system. On one hand, we permit validity tests by means of a spherical restriction. On the other hand, from the view point of the user, we try to work in the maximal cube that belongs to the interior of the sphere, i.e. to the set

$$\bar{D}E^3(m) = \{(x,y,z) \in DE^3 \cap M_1 \mid max\{|x|,|y|,|z|\} < \frac{1}{3}\sqrt{3} \cdot 10^{m/2}\}.$$

The upper bound y_{max} of the number representation reduces to $\bar{y}_{max}$. Defining

$$\bar{y}_{max} = 0.5 \cdot 10^{m/2}$$

we get a better way to restrict the input data. Indeed, we still take the sign of the components into consideration, since the objects can be represented in all three components independently with a maximal extension of $2\bar{y}_{max} = 10^{m/2}$ and a separation precision of $y_{min} = 10^{-m/2}$. Using six significant digits corresponds to the maximal extension of the natural scale units from about 10 meters up to a detail of $\frac{1}{100}$ millimeter.

We continue the basic idea of Lemma 6.4 to compute scalar products successively and sufficiently correctly without a number domain overflow.

Definition 6.5. *Let* $M_2 = \{(x,y,z) \in E^3 \mid x^2 + y^2 + z^2 < \frac{1}{3})10^m\}$. *The set* $DE^3(m) \cap M_2$ *is called a* **doubly restricted discretization** *of* E^3.

The maximal cube $\tilde{D}E^3(m)$ belonging to the doubly restricted discretization is defined by

$$\tilde{D}E^3(m) = \{(x,y,z) \in DE^3 \cap M_2 \mid max\{|x|,|y|,|z|\} < \frac{1}{3}10^{m/2}\}.$$

The upper bound y_{max} of the number domain reduces to $\tilde{y}_{max} = \frac{1}{3}10^{m/2}$. Obviously, the inclusions

$$\tilde{D}E^3(m) \subset \bar{D}E^3(m) \subset DE^3(m)$$

are valid.

Lemma 6.6. *Let $a, b \in \tilde{D}E^3(m)$. Then their scalar product (a, b) is computable in $DR^1(2m)$ without rounding errors and $|(a, b)| < \frac{1}{3}10^m$. Their vector product $[a, b]$ is computable in $DR^3(2m)$, and representable in $\tilde{D}E^3(2m)$ analogously.*

Proof. The scalar product (a, b) is computable in $DR^1(2m)$ because of $\tilde{D}E^3(m)$ and Lemma 6.4. From $|a_i|, |b_i| < \frac{1}{3}10^{m/2}$, $(i = 1, 2, 3)$ it follows that $|(a, b)| < \frac{1}{3}10^m$. According to the first component p_1 of the vector product $[a, b]$, we have

$$p_1 = a_2b_3 - a_3b_2 < \frac{2}{9}10^m < \frac{1}{3}10^m$$

and the analogous result $p_2, p_3 \leq \frac{1}{3}10^m$. The vector product $p = [a, b]$ is computable in $DR^3(2m)$ because of Lemma 6.4, then p can be represented exactly in $\tilde{D}E^3(2m)$. □

In addition to our correct scalar product, it is still necessary to define overflow domains describing difference vectors of E^3. This is done by additional restrictions on the domains mentioned above, namely the first octant. Then we denote the domain notations by a leading symbol +. Figure 6.1 indicates a projection of the described domains. The unused parts of

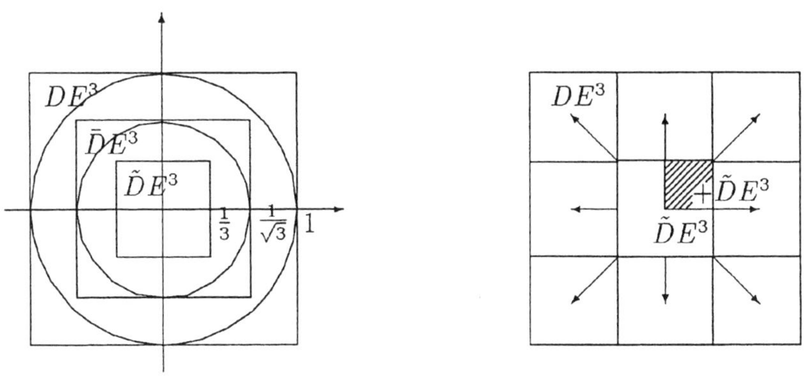

Figure 6.1. Restricted domains: their additional positioning and overflow domains.

$DE^3(m)$ during object definition permit organization of overflow domains of translations. They may also be used for additional positioning to store other objects or a different view of the object being defined. If we work on $\tilde{D}E^3$, then twenty six adjacent and homogeneous domains are possible.

$DE^3(m)$ also permits allocation of rotations placement as an overflow domain of $+\tilde{D}E^3(m)$ because the diagonal length $1\frac{1}{3}\sqrt{3}\cdot 10^{m/2}$ of $+\bar{D}E^3(m)$ is smaller than the edge length $\frac{2}{3}10^{m/2}$ of the single overflow domain.

The doubly restricted discretization of E^3 may be generalized to E^4. We need this generalization for considering scalar products with four-dimensional vectors and for receiving analogous assertions to describe the geometry by means of homogeneous coordinates.

Definition 6.7. *Let $M_2 = \{u \in E^4 | (u,u) < \frac{1}{4}10^m\}$. The set $DE^4 \cap M_2$ is called a doubly restricted discretization of E^4 with its maximal cube domain*

$$\tilde{D}E^4 = \{u \in DE^4 \cap M_2 \mid max\{u_1, u_2, u_3, u_4\}| < \frac{1}{4}10^{m/2}\}$$

and $\tilde{y}_{max} = \frac{1}{4}10^{m/2}$.

The assertion of Lemma 6.6 and its proof also applies to the scalar product.

Lemma 6.8. *The scalar product (a,b) of $a, b \in \tilde{D}E^4(m)$ is correctly computable in $DR^1(2m)$ and $|(a,b)| < \frac{1}{4}10^m$.*

6.2. Representation of a Polynomial Equation

Not only correct points are needed for describing boundary representations, but also correct equations of curves and surfaces are needed.

A Polynomial Equation without Rounding

Firstly, we consider the correctness of one equation, i.e., a surface in E^3 or a curve in E^2. A curve in E^3 will be treated later.

Equation parameters have an algebraic nature. But it is advantageous to define and modify geometric objects by means of geometric parameters. The relations between algebraic and geometric parameters are of fundamental importance to the organization of the modeling process.

There are many different ways to describe surfaces by means of geometric properties, for example by points, tangents, curvatures, and so on. In the following context we restrict ourselves to describing the surface or the curve in E^2 by defining points (geometric parameters) on them. Also, we are interested in a transformation of the geometric parameters (the defining points) into algebraic ones (the curve parameters), according to the precision of $DE^3(m)$. Descriptions by means of other geometric parameters can be reduced to these definitions. For most purposes, it is sufficient to consider polynomial equations in three (or four) variables up to polynomial degree three.

Let us consider those polynomial equations in a recursive form, so that we can later describe the precision estimates by means of scalar products only.

Definition 6.9. *Let $r = (x, y, z)^T \in E^3$. A real function $p = p(r, 0)$ denotes a polynomial of degree 0. The real functions $p_1(r, n-1)$, $p_2(r, n-1)$ and $p_3(r, n-1)$ are to be polynomials of the degree $n-1$ with $(p_1, p_2, p_3)^T \neq 0$ if $r \neq 0$ and $q(n) \in R^1$. Then*

$$p(r, n) = xp_1(r, n-1) + yp_2(r, n-1) + zp_3(r, n-1) + q(n)$$

is called a **polynomial of degree** *n.*

At first we consider the precision of algebraic parameters of a linear equation. The linear equation

$$-q = p_1 x + p_2 y + p_3 z$$

with $p = (p_1, p_2, p_3)^T \in E^3, r = (x, y, z)^T \in E^3$ and $q \in R^1$ arises by equating a polynomial of degree 1 to zero. This equation can only be correctly approximated by rounding the algebraic parameters p_1, p_2, p_3 and q if we use the same rounding error of both sides by rounding the equation corresponding to its worst case on both sides. Therefore it is obvious to distinguish between the precision of the special parameter q and the parameters p_1, p_2, p_3. Moreover, the precision of q has to be chosen depending on the other three parameters and the precision of r.

The algebraic parameters of such a linear equation can be defined in a geometric way by any of following:

- a fixed point and the normal vector
- three linearly independent points
- a fixed point and two directions,

thus defining a plane. The first two cases are sufficient to explain the precision estimate.

The first case is very easy. If $p, r \in \bar{D}E^3(m)$, then $q = p^T r$ is computable in $DR(2m)$ because of Lemma 6.6.

In the second case, three points $u, v, w \in E^3$ are given and the plane can be defined using a determinant

$$\begin{vmatrix} x - u_1 & y - u_2 & z - u_3 \\ v_1 - u_1 & v_2 - u_2 & v_3 - u_3 \\ w_1 - u_1 & w_2 - u_2 & w_3 - u_3 \end{vmatrix} = 0.$$

An equivalent form is

$$\begin{vmatrix} x & y & z \\ v_1 - u_1 & v_2 - u_2 & v_3 - u_3 \\ w_1 - u_1 & w_2 - u_2 & w_3 - u_3 \end{vmatrix} = \begin{vmatrix} u_1 & u_2 & u_3 \\ v_1 - u_1 & v_2 - u_2 & v_3 - u_3 \\ w_1 - u_1 & w_2 - u_2 & w_3 - u_3 \end{vmatrix}.$$

On the left side, the description of the normal vector p of the plane is given as the vector product $p = [u - v, w - u]$. If $u, v, w \in +\tilde{D}E^3(m)$, then the difference vectors $u - v, w - u$ belong to $\tilde{D}E^3(2m)$ because of Lemma 6.6. The right hand side of the equation can be written as the scalar product $u^T p$ between the given point u and the normal vector p. It is computable in $DR^1(3m)$ because no number domain has to be changed.

In the second case, difficulties in computing and storing the equation arise if only simple and double precision are possible according to the precision of $DE^3(m)$. But threefold precision is sufficient and may be simulated by means of special programs.

Polynomial equations of degree two can be described analogously by means of a determinant. If we fix the quadratic equation

$$ax^2 + by^2 + cz^2 + dxy + eyz + fzx + gx + hy + kz + q = 0$$

with the variables x, y, z by nine special points $(x_1, y_1, z_1) \cdots (x_9, y_9, z_9)$, we obtain the linear system

$$\begin{array}{ccccccccc} ax_1^2 & + & by_1^2 & + & cz_1^2 & + & \cdots & = -q \\ \vdots & & \ddots & & \vdots & & \vdots & = -q \\ ax_9^2 & + & by_9^2 & + & cz_1^2 & + & \cdots & = -q \end{array}$$

for the parameters of the quadratic equation. If the coefficient matrix is nonsingular then the parameters on the left hand side can be expressed by means of determinants according to Cramer's rule. Exchanging the parameters in the quadratic equation by these determinants, and putting the first given point into the equation, we can eliminate the parameter q.

Then the quadratic equation takes on the form

$$\begin{vmatrix} x^2 - x_1^2 & \cdots & xy - x_1y_1 & \cdots & z - z_1 \\ x_2^2 - x_1^2 & \cdots & x_2y_2 - x_1y_1 & \cdots & z_2 - z_1 \\ \vdots & \ddots & \vdots & \ddots & \vdots \\ x_9^2 - x_1^2 & \cdots & x_9y_9 - x_1y_1 & \cdots & z_9 - z_1 \end{vmatrix} = 0,$$

only dependent on geometric parameters. In this form, the desired left hand side parameters are expressed by the nine eight-dimensional subdeterminants of the first row. Computations of these determinants, without rounding, is very expensive.

The expense goes up depending on a larger degree n of the desired polynomial equation. If we multiply the geometric input data by $10^{-m/2}$, then the given points are integers and so are the determinants. Using Gaussian elimination, and avoiding division by sufficient scaling of the rows, we only need one division at the end of Gaussian elimination which is computed without any rounding. But the disadvantage lies in the exponentially growing number of digits with respect to the degree n of the polynomial equation. If we compute the determinants according to their definitions by means of permutations, then the run-time increases exponentially with respect to n.

Therefore, it seems to be impossible to define algebraic parameters of a polynomial equation with a degree of more than one without rounding.

Floating Point Representation of a Polynomial Equation

The negative statement about representing the algebraic parameter without rounding leads to the question of a floating point representation. We will see that it is not only promising but also practicable. Strictly speaking, the results of polynomials with a high degree may be bad in a geometric sense. It is necessary to ask how many digits have to be used for sufficient geometric properties. Curious effects already arise by rounding parameters of linear equations.

To understand such effects, let us consider an example of linear equations. This example will show that a very small perturbation of the algebraic parameters sometimes supplies results in a large perturbation of the geometric parameters which defined the linear equation. It is sufficient to take a linear equation in E^2 and to consider the corresponding straight line.

Example 6.10. The straight line $g_1 \subset E^3$ is given by two points

$$P_1(400., 500.) \text{ and } P_2(400.001, 900.) \text{ of } DE^2(6).$$

The exact linear equation reads as follows

$$-400.x + 0.001y + 159999.5 = 0..$$

Now we round the absolute component to six significant digits and compensate the induced error by a translation of the straight line into the direction of the y-axis, i.e. we get

$$-400.x + 0.001\bar{y} + 160000. = 0.$$

and $\bar{y} = y - 500.$. The points P_1 and P_2 are translated to the points $Q_1(400., 0.)$ and $Q_2(400.001, 400.)$, see Figure 6.2.

At first this seems very undesirable because of the large geometric modification. But this modification into the direction of the y-axis represents a sufficiently small common difference. The orthogonal distance between the old and new straight line is smaller than 0.0025.

The so called backward analysis, taken in the example above, leads to the large error $DELTA = 500$. Concerning the geometric properties of the whole straight line, the rounding yields the geometric distance $DELTA = 0.0025$. We need a special notion to distinguish between different views in a geometric-topologic sense.

Two operators given by the parameter vectors a and $\bar{a} \in R^l$, map the two points $x, \bar{x} \in E^n$ into R^1, denoted by the symbol $\circ$. Any norm in E^n

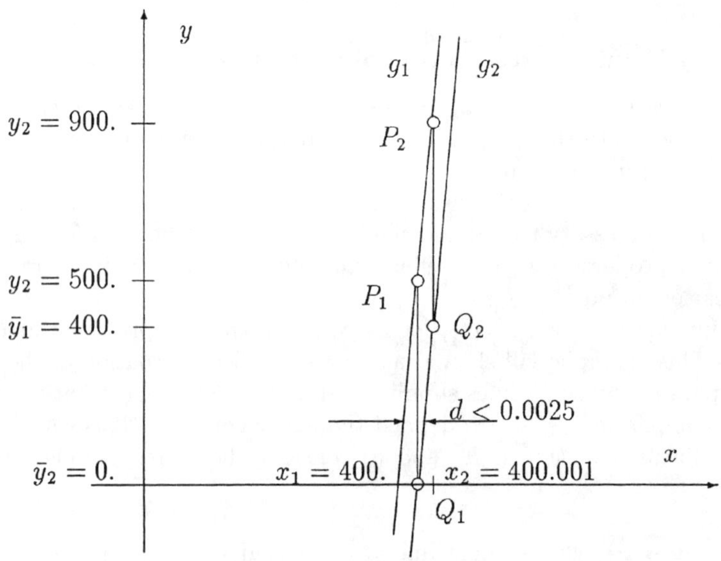

Figure 6.2. By rounding an algebraic coefficient, large changes of geometric parameters (points) defining a straight line can result, and even with a geometrically suitable result.

is denoted by $\| \,.\, \|$. The following defines how two geometric forms are the same in the sense of the representation precision.

Definition 6.11. *If there exists a $\bar{x}$ for all $x \in M_i \cap DE^n(m)$ and visa versa such that*

$$\| x - \bar{x} \| < EPS = 0.5 \cdot 10^{-m/2},$$

then the operators a and $\bar{a}$, their corresponding equations

$$a \circ x = 0 \quad \text{and} \quad \bar{a} \circ \bar{x} = 0$$

and geometric forms in E^n are called **geometrically suitable** *over the set $DE^n(m)$, with its corresponding representation precision EPS.*

The two approximation examples in Figure 6.3 demonstrate that each suitable approximation curve f_a approaches the given curve f in each of its points with a sufficiently small distance and visa versa. Now we want to use this definition to describe the rounding

$$rd(a_i) = \bar{a}_i \quad (i = 1, 2, ..., n)$$

of n algebraic parameters a_i to $\bar{a}_i$. The definition shows that a rounding is geometrically suitable if it yields a sufficiently small difference of the desired geometric form.

Lemma 6.12. *Let $u \in \tilde{D}E^4(m)$, $a \in E^4$ and $rd(a) = b \in DE^4(m)$. Then the rounded equation $u^T b = 0$ is geometrically suitable to the equation $u^T a = 0$ according to the Euclidean norm $\| \,.\, \|_2$.*

Proof. The rounding error of one component $a_i (i = 1, 2, 3, 4)$ to a floating point number with m significant digits can be expressed by

$$b_i = rd(a_i) = a_i(1 + EPS_i)$$

Figure 6.3. A geometrically suitable and a non-suitable approximation.

and the relative precision EPS_i is restricted by $|EPS_i| \leq 0.5 \cdot 10^{-m}$. It yields the following non-perturbed and perturbed equations

$$\begin{array}{l} a_1u_1 + a_2u_2 + a_3u_3 + a_4u_4 = 0 \\ a_1(1+EPS_1)u_1 + a_2(1+EPS_2)u_2 + a_3(1+EPS_3)u_3 + a_4(1+EPS_4)u_4 = 0. \end{array}$$

Now we do not interpret the perturbation $(1+EPS_i)$ in the second equation as parameter differences, but as a difference of the components of the vector u. Accordingly, we define a new variable vector $\bar{u}$ in the second equation. Dividing the two equations by the norm of the parametric vector we get

$$\frac{a^T u}{\| a \|} = 0 \quad \text{and} \quad \frac{a^T \bar{u}}{\| a \|} = 0$$

with $\bar{u}_i = (1+EPS_i)u_i \quad (i = 1, 2, 3, 4)$. The distance t between a point $\bar{u}$ and the hyper plane of the first equation yields

$$t = \frac{|a^T \bar{u} - a^T u|}{\| a \|} = \frac{1}{\| a \|} |\sum_{i=1}^{4} a_i u_i EPS_i|$$

and the estimate

$$t \leq \frac{eps}{\| a \|} a^T u \qquad \textit{with} \qquad eps = maxEPS_i \leq 0.5 \cdot 10^{-m}.$$

From $\frac{a_i}{\|a\|} \leq 1 \ (i = 1, 2, 3, 4)$, the triangle inequality and $u \in \tilde{D}E^4$ yield

$$t \leq 0.5 \cdot 10^{-m} |\sum_{i=1}^{4} u_i| \leq 0.5 \cdot 10^{-m} 4 \cdot 0.25 \cdot 10^{m/2}$$

$$t \leq 0.5 \cdot 10^{-m/2} = EPS. \ \square$$

Treating $u_4 = 1$ as a constant, Lemma 5.8 shows that in the discretization $\tilde{D}E^3(m)$ of E^3, whose scalar products are restricted by $10^{-m}/4$, the plane equation is geometrically suitable using simple precise rounding. This is a very practical result.

Taking a domain without restrictions, the estimate of t could only yield $t \leq 4 \cdot EPS$. This result indeed does not lead to geometrically suitable approximations but it is still useful. One only has to pay attention to determine the conformity of planes by means of a four times larger tolerance than the drawing precision EPS. The Example 6.10 already shows that a tolerance of more than EPS of any point and its map can arise if we do not restrict the domain. Nevertheless, the rounding result is geometrically suitable.

This practical way of defining linear equations in the sense of geometrically suitable approximations becomes problematic when using equations with polynomials of higher degree. The following example shows the rounding of algebraic parameters of one equation in E^2, i.e., of a curve, that is not a suitable rounding in the geometric sense.

Example 6.13. Determine a quadratic equation $y = x^2 + ax + b$ that is satisfied by the two points $P_1(400., 0.)$ and $P_2(400.002, 0.)$ of $DE^2(6)$.

The points P_1 and P_2 can be interpreted as the zeroes of the equation

$$y = (x - 400.)(x - 400.002)$$

i.e.,

$$r_1 : \qquad y = x^2 - 800.002x + 160000.8.$$

If we round the algebraic parameters with respect to the precision $DR(6)$ to six significant digits, we obtain

$$r_2 : \qquad \bar{y} = x^2 - 800.002x + 160001..$$

Obviously $\bar{y} = y + 0.2$ and the rounding corresponds to a translation of the curve by 0.2 units in the direction of the y-axis. Now the distance d of the extreme point $S_1 = (x_0, y_0)$ of the curve r_1 to r_2 is interesting, see Figure 6.4.

On one hand, the roundings of an algebraic coefficient sometimes produce small tolerances of the defining geometric parameters (here the points

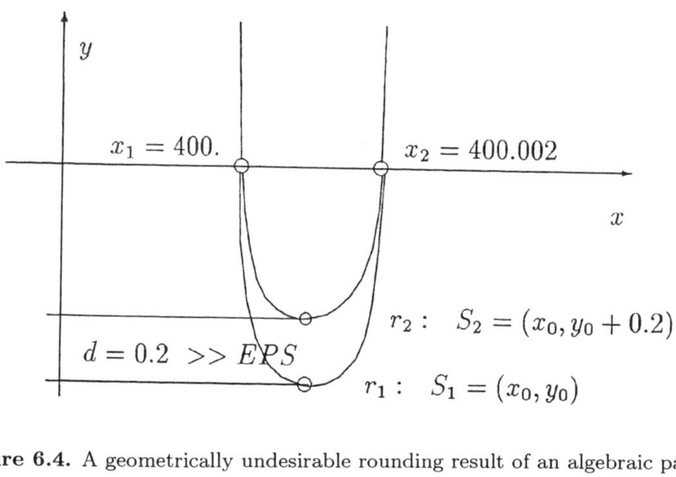

Figure 6.4. A geometrically undesirable rounding result of an algebraic parameter.

$(x_1, 0.)$ and $(x_2, 0.)$), but on the other hand they may produce large geometric errors. It is the distance to the point $S_2 = (x_0, y_0 + 0.2)$

$$d = \| (x_0, y_0) - (x_0, y_0 + 0.2) \|_2 = 0.2 = 200EPS >> EPS$$

and is not geometrically suitable in $DE^2(6)$. The rounding is 200 times worse than justifiable.

The example shows that single precision in the numbers representing algebraic parameters of an equation with a higher degree is not sufficient to find the suitable rounding. The question arises, how many more digits are necessary to define a geometrically suitable rounding. The following example suggests the trend.

Example 6.14. Let us consider all points $(x, y) \in DE^3(6)$ which satisfy

$$y = x^n - a \quad (n = 1, 2, 3, ...).$$

We define the parameter a by a point P that corresponds to the zero of the equation, for example $P(500.001, 0.)$.

The parameter a can be computed as $a = x^n$. Then the necessary number of significant digits of a without rounding is equal to $6n$ since we have taken the origin P from a centered representation with six digits, i.e. from $DE^3(m)$. Now we want to approximate the y-value by one, two, three and sixfold precision of the $m = 6$ digits, i.e., $rd(y) \in DE^3(k \cdot 6)$ and $k = 1, 2, 3, 4$ (see Table 6.1). The rounding value corresponds to the translation value parallel to the y-axis. As in Example 6.13, we consider the lowest point in y-direction of the input equation and the rounded equation depending on the digit factor of the assigned decimal number of a.

The equation is geometrically suitable according to the representation precision $0.5 \cdot 10^{-3}$ of $DE^3(6)$ with its centered and fixed-point numbers

n	$m = 1 \cdot 6$	$m = 2 \cdot 6$	$m = 3 \cdot 6$	$m = 4 \cdot 6$
1	$0.5 \cdot 10^{-3}*$	$0.5 \cdot 10^{-3}$	$0.5 \cdot 10^{-3}$	$0.5 \cdot 10^{-3}$
2	$[0.5 \cdot 10^0]$	$0.5 \cdot 10^{-6}!$	$0.5 \cdot 10^{-6}$	$0.5 \cdot 10^{-6}$
3	$0.5 \cdot 10^3$	$0.5 \cdot 10^{-6}!$	$0.5 \cdot 10^{-9}$	$0.5 \cdot 10^{-9}$
4	$[0.5 \cdot 10^6]$	$[0.5 \cdot 10^0]$	$0.5 \cdot 10^{-6}$	$0.5 \cdot 10^{-12}$
5	$0.5 \cdot 10^9$	$0.5 \cdot 10^3$	$0.5 \cdot 10^{-3}$	$0.5 \cdot 10^{-9}$
6	$[0.5 \cdot 10^{12}]$	$[0.5 \cdot 10^6]$	$[0.5 \cdot 10^0]$	$0.5 \cdot 10^{-6}$

Table 6.1. Representation precision of y (Example 6.14) independent of the number of digits m of the parameter a and the exponent n.

only above the horizontal lines in Table 6.1. On the other hand, the marked results are not geometrically suitable to $DE^3(6)$. Linear geometry with single precision (see *) has a special significance if only single and double precision are available because double precision can be taken as a reserve. Quadratic and cubic equations (splines) already have to be represented by means of double precision (see !), and their computing reserve must be higher than two.

Table 6.1 does not give a general answer as to how many digits are sufficient for describing geometrically suitable equations of degree n. Let $p(r,n)$ with $r = (x,y,z) \in DE^3(m)$and $n \in G$ be a polynomial so that we write

$$p(r,n) = a_1x^n + a_2y^n + a_3z^n + \cdots \quad .$$

The domain $DE^3(m)$ is a fixed-point domain according to its definition with the absolute error $EPS = 0.5 \cdot 10^{-m/2}$ (drawing precision). If we consider a point $r = (x,y,z)$ of the desired polynomial equation, then it is sufficient to describe geometrically suitable approximations by using differences which do not exceed EPS. We incorporate these differences of each term of $p(r,n)$ into the corresponding coefficients in forms of a relative error because we want to estimate their sensitivity. The first term yields

$$a_1(x + EPS)^n = (1 + \frac{EPS}{x})^n a_1x^n = (1+E)a_1x^n$$

with

$$\frac{EPS}{x} \leq 0.5 \cdot 10^{-m} \quad (x \geq 0)$$

and

$$\begin{aligned} E &\leq -1 + (1 + 0.5 \cdot 10^{-m})^n \\ &= -1 + (1 + ... + 0.5^n \cdot 10^{-mn}). \end{aligned}$$

It yields the smallest relative perturbation of $0.5^n \cdot 10^{-mn}$. Analogous statements hold for the remaining terms. Then it is possible that the mn-th significant digits of the coefficients are utilized. But more than mn significant digits are not necessary to represent the equations.

Lemma 6.15. *The coefficients of the polynomial $p(r,n)$ with degree n and $r = (x,y,z) \in DE^3(m)$ are rounded to nm significant digits corresponding to the relative precision $EPS_{rel} = 0.5^n \cdot 10^{-mn}$. These roundings transform the equation*

$$p(r,n) = 0$$

in a geometrically suitable way according to the maximum norm $\| \, . \, \|_\infty$ of E^3. A rounding with fewer digits may violate this geometric result.

Let us go back to the Table 6.1 of Example 6.14. We could notice that in this special case the numbers on the horizontal lines represent geometrically suitable approximations depending on the degree and the number of digits. Sometimes Lemma 6.15 changes these lines disadvantageously with respect to the number of digits. But it also shows that the necessary (neglecting 0.5^n) number $\bar{m} = mn$ of significant digits representing the coefficients leads to geometrically suitable approximations. Therefore, in order for the algebraic parameter to be geometrically suitable, we need about the following:

linear equation	-	single precision
quadratic equations	-	double precision
cubic equation	-	threefold precision

and so on, if the points of E^3 are defined in single precision of centered fixed-point representations.

6.3. The Solution of a Floating Point Equation

In calculating boundary representations of geometric objects, implicit equations and systems of equations, as well as their solutions, are needed. If a solution to these equations exists, it can be obtained in one of two different ways: firstly, by equivalence transformations into simplified equations or secondly, by iterative methods. In this chapter, we are only interested in the first case, which is the analytical way.

Let us consider the calculation of the function value $y = f(x)$ of a sufficiently often differentiable function f. Let the result $\tilde{y} = y + \delta_y$ of the calculation be perturbed by the term $\tilde{x} = x + \delta_x$ and by the approximating computation $\tilde{y} = \tilde{f}(\tilde{x})$.

Estimating the error $\delta_y = \tilde{f}(\tilde{x}) - f(x)$ in the sense of forward analysis, the value $\tilde{f}(\tilde{x})$ is interpreted as a modified exact computation $\tilde{f}(\tilde{x}) = (1 + \varepsilon_f) f(\tilde{x})$. The perturbation ε_f is restricted to $\| \varepsilon_f \| < E_f$ by a given bound E_f according to the precision of the function value.

Generating the Taylor series of $f(\tilde{x}) = f(x + \delta_x)$ at the argument x, we get the *absolute total error* δ_y of first order,

$$\delta_y \approx \varepsilon_f f(\tilde{x}) + \delta_x f'(\tilde{x}) \quad \textit{with} \quad \| \varepsilon_f \| < E_f, \| \delta_x \| < \Delta_x,$$

and the *relative total error* $\delta y / f(\tilde{x})$ of first order,

$$\delta y / f(\tilde{x}) \approx \varepsilon_f + \frac{\tilde{x} f'(\tilde{x})}{f(\tilde{x})} \varepsilon_x \quad \textit{with} \quad \| \varepsilon_f \| < E_f, \| \varepsilon_x \| < E_x.$$

Here, the Taylor series was truncated after the second term. That is sufficient for our purposes. The terms

$$\varepsilon_f f(\tilde{x}) \quad and \quad \varepsilon_f$$

are called *absolute produced error* and *relative produced error* , respectively. They are only caused by the approximated value computation. The differences

$$f(\tilde{x}) - f(x) \quad and \quad 1 - f(x)/f(\tilde{x})$$

are called *absolute propagated error* and *relative propagated error*, respectively.

Increasing factors of the absolute error δ_x and the relative error ε_x appear in the total error. Those factors

$$acond(f) = f'(\tilde{x}) \quad or \quad cond(f) = \frac{\tilde{x}f'(\tilde{x})}{f(\tilde{x})}$$

are called absolute or relative *condition numbers* of the problem $y = f(x)$. If their absolute value is less than or equal to 1, then the multiplied error is in the same order of magnitude as the input error. In this case one calls the problem $y = f(x)$ *well-conditioned problem* and otherwise the problem is called an *ill-conditioned problem.*

If the argument of the problem $y = f(x)$ is an element of $R^n \quad (n > 1)$, then equivalent notations can be used. The only difference is the following one. A Taylor series of several variables has to be used instead of the Taylor series for a function of a single variable. So the condition numbers of the addition, subtraction and multiplication of two numbers $x_1, x_2 \in R^1$ can be defined by their incorrect arguments $\tilde{x}_1$, $\tilde{x}_2$. High absolute condition numbers are reached in divisions with a divisor of a low absolute value. High relative condition numbers arise in subtractions of nearly equal numbers. In this case significant digits are reduced.

These unfavorable conditions should be avoided in algorithms. An algorithm is a *stable algorithm* if the condition numbers of the arguments and all the partial results are well-conditioned. Otherwise, it is an *unstable algorithm* . We should give some examples of condition numbers: The function $f(x) = x^n$ with rational exponent $n \neq 0$ has the absolute condition $acond(x^n) = n\tilde{x}^{n-1}$ and the relative condition $cond(x^n) = n$. The function $f(x) = sin(x)$ has the condition $acond(sin(x)) = cos(\tilde{x})$ or $cond(sin(x)) = \tilde{x}cot(\tilde{x})$.

Intensive computation is commonly called for in the modeling process. Sometimes thousands of problems of one and the same type have to be calculated to define a special object. If only one of those problems was unstable, the object produced can be unrecognizably deformed. Therefore, it is necessary to remark on how to improve the condition numbers by considering the number of significant digits of the equation parameters. The rounding of the two sides of an equation with real parameters into floating-point numbers leads to a new equation. This kind of equality is also an equivalence relation and one can work with this equation as usual. Relative condition numbers of intermediate results indicate how the necessary numbers of significant digits are to be modified.

Problem 6.16. *The solution x in the form of a floating-point representation $rd_m(x)$ of the quadratic equation*

$$x^2 + 2px + q = 0 \quad with \quad p^2 \geq q,$$

is desired, such that the parameters are given by $rd_m(r)$,$rd_{2m}(s)$ and $rd_{3m}(t)$.

Putting the parameter p and q into the corresponding solution formula, significant digits can be removed in the differences. Therefore one first computes the well-conditioned zero, according to the known solution formula for quadratic equations, and after that the other zero by means of the Vieta rule

$$\begin{aligned} x_1 &= -p - sign(p)\sqrt{p^2 - q} \\ x_2 &= q/x_1. \end{aligned}$$

In crossing to the floating-point numbers, the first equation yields

$$\begin{aligned} rd_m(x_1) &= rd_m(-p - sign(p)\sqrt{p^2 - q}) \\ &= rd_m(rd_m(-p) - sign(p) rd_m\sqrt{p^2 - q}). \end{aligned}$$

The relative condition number of the square root computation $\sqrt{u}$ is equal to $1/2$ and it is 2 for the quadratic u^2. The first zero will be expressed approximately by

$$\tilde{x_1} = rd_m(rd_m(x_1) \approx (rd_m(-p) - sign(p) rd_m(\sqrt{rd_{2m}((rd_m p)^2 - rd_{2m}(q)}))).$$

A possible cancellation of significant digits does not play a role in the solution $\tilde{x}_1$ if the difference $p^2 - q$ was computed with double precision, as demanded by x_1. The relative condition number of a division has absolute value equal to 1, and therefore

$$\tilde{x}_2 = rd_m(x_2) = rd_m(q/x_1) \approx rd_m(\frac{rd_m(q)}{rd_m(x_1)}).$$

These floating-point solutions demand special properties for the parameter p and q, according to their precision. To compute x_1 as written above with a relative precision of $eps = 10^{-m}$ then the parameter p has to take at least the same precision and the parameter q must be double precision relative to x_1, i.e., $eps = 10^{-2m}$.

Of course, in the field of modeling, it is not as interesting to know the relative precision of x_1 and x_2 as the absolute precision. The variable x has a geometric interpretation in this context, i.e., it represents a component of a point of Euclidean space E^3. The discretization $DE^3(m)$ of this space is based on a centered fixed-point representation with m digits. Each fixed-point domain with m digits can be embedded in floating-point numbers with m significant digits.

In such embedded domains of fixed numbers, the relative error is equal to the absolute error. The considerations of the relative precision of the solutions and the attached precision of the equation parameters, by means of the formulas mentioned above, are valid in this embedding case in order to get the well-defined absolute error of $x \in DE^3(m)$.

Calculations of fixed-point numbers by means of adapted floating-point numbers have the advantage that the intermediate results do not have to be described by an expensive normalization. We want to follow this idea in the case of polynomial equations of third and fourth degree.

Problem 6.17. *Compute all the real solutions x in the form of the floating-point representation $rd_m(x)$ of the cubic equation*

$$x^3 + rx^2 + sx + t = 0$$

If we substitute $x = y - \frac{r}{3}$ and define $p = \frac{1}{3}(s - \frac{r}{3})^2$ and $q = \frac{1}{2}(\frac{2r^3}{27} - \frac{rs}{3} + t)$, then the cubic equation to be solved transforms into the so-called reduced form

$$y^3 + 3py + 2q = 0.$$

Let D and $u_{1/2}$ be defined by

$$D = q^2 + p^3 \quad \textit{and} \quad u_{1/2} = \sqrt[3]{-q \pm \sqrt{D}}.$$

Then the solutions of the reduced equation are the following three

$$\begin{aligned} y_1 &= u_1 + u_2 \\ y_{2/3} &= -\frac{u_1 + u_2}{2} \pm \frac{u_1 - u_2}{2} i\sqrt{3}. \end{aligned}$$

The possible real solutions of those equations differs from one another by the sign of the discriminant D (see Table 6.2).

The special case, $D = 0$ and $p = 0$, leads to $q = 0$, and therefore it can only be met in the form of the trivial cubic equation $y^3 = 0$ whose solution

number of real terms	case	real solutions
1	$D > 0$	y_1
2	$D = 0$	$y_1, y_2 = y_3$
3	$D < 0$	$y_1,\ y_2,\ y_3$

Table 6.2. All real terms of solutions y_i of the reduced cubic equation.

$y_{1/2/3}$ are zeros.Those known trivial equations can be arranged again into the cases shown in Table 6.2

$$\begin{array}{rcllll} 0 &=& y^3 & \textit{case} \quad D = 0 & \textit{with} \quad p, q = 0 & \\ 0 &=& y^3 + 3py & \textit{case} \quad D < 0 & \textit{with} \quad p = 0 & \textit{and} \quad q < 0 \\ 0 &=& y^3 + 2q & \textit{case} \quad D < 0 & \textit{with} \quad q = 0 & \textit{and} \quad q < 0. \end{array}$$

In the case $D > 0$, u_1 and u_2 and their radicands are real. The real solution $y_1 = u_1 + u_2$ is formed into the associated floating-point representation

$$\begin{array}{rcl} rd_m(y_1) & \approx & rd_m(rd_m \sqrt[3]{rd_{3m}(rd_{3m}(-q) + rd_{3m}D)} \\ & & +rd_m \sqrt[3]{rd_{3m}(rd_{3m}(-q) + rd_{3m}D)}) \end{array}$$

with

$$D \approx \sqrt[3]{rd_{6m}((rd_{6m}(rd_{3m}q)^2) + rd_{6m}((rd_{2m}p)^2))}.$$

If the intermediate results are computed using the precision in the formula, a good estimate of the solution y_1 can be reached, corresponding to its fixed-point precision (drawing precision of E^3). Of course the parameter p has to be defined in triple precision and q in double precision.

In the case $D = 0$, it holds that $u_1 = u_2 = \sqrt[3]{-q}$, with the single solution $y_1 = 2u_2$ and the double solution $y_2 = y_3 = -u_1$. The floating-point representation of the two solutions y_1, y_2 yield

$$\begin{array}{rcl} rd_m(y_1) & \approx & rd_m \sqrt[3]{rd_{3m}(-q)} \quad \textit{and} \\ rd_m(y_2) & \approx & rd_m(2rd_m(y_1)). \end{array}$$

The case $D < 0$ is a little complicated. The intermediate results are not always real numbers because of the negative sign of the discriminant. If we write these complex values in a trigonometrical form, then real results can be reached in a common real form by means of the Moivre formula. Let $r = \sqrt{-p^3}$ and $\cos\varphi = -\frac{q}{2r}$. Then we get the three real solutions $(i = 1, 2, 3)$

$$y_i = 2\sqrt[3]{r}\cos(\frac{\varphi}{3} + (i-1)120^o).$$

The computation of φ needs the function arccos. This function does not permit a simple floating-point form of these equations. Therefore it is better to determine the real solution, in the third case $D < 0$, by computing with complex arguments. In this case, the two possible radicands $-q + \sqrt{D}$ and $-q - \sqrt{D}$ and their roots u_1 and u_2 are conjugate complex numbers. Then we can suppose that $u_1 + u_2$ is a real number and $u_1 - u_2$ has only a complex part. The associated floating-point representation of the three real solutions y_1, y_2 and y_3 yields an analogous representation of y_1 as in the first case $D > 0$. The required precision of the equation parameters p and q are double and triple, respectively, just as in the other cases.

Problem 6.18. *Compute all real solutions x in the floating-point representation $rd_m(x)$ of the fourth degree equation*

$$x^4 + ax^3 + bx^2 + cx + d = 0,$$

such that the parameters are given by $rd_m(a)$, $rd_{2m}(b)$, $rd_{3m}(c)$ and $rd_{4m}(d)$.

The substitution $x = y - \frac{1}{4}a$ and the definitions $e = b$, $f = \frac{1}{8}a^3 + \frac{1}{2}ab + c$ and $g = -\frac{3}{256}a^4 + \frac{1}{16}a^2b - \frac{1}{4}ac$, transforms the equation of fourth degree into the reduced equation

$$y^4 + ey^2 + fy + g = 0,$$

whose four solutions $y_1, \cdots, y_4$ can be computed by means of the equations

$$\begin{aligned} 2y_1 &= \sqrt{z_1} + \sqrt{z_2} + \sqrt{z_3} & \qquad 2y_2 &= \sqrt{z_1} - \sqrt{z_2} - \sqrt{z_3} \\ 2y_3 &= -\sqrt{z_1} + \sqrt{z_2} - \sqrt{z_3} & \qquad 2y_4 &= -\sqrt{z_1} - \sqrt{z_2} + \sqrt{z_3} \end{aligned}$$

where the numbers z_1, z_2, z_3 can be taken from the roots of the cubic equation

$$z^3 + 2ez^2 + (e^2 - 4g)z - f^2 = 0.$$

We assume the precision of the reduced equation of fourth degree as 10^{-4m}, and the representation of its coefficients as $rd_{2m}(e), rd_{3m}(f)$ and $rd_{4m}(g)$. The absolute precision 10^{-m} of the solutions y_i $(i = 1, \cdots, 4)$ in floating-point representation $rd_m(y_i)$ demand that the roots of z_i $(i = 1, 2, 3)$ be in a double precision representation, i.e., $rd_{2m}(z_i)$. The coefficient precision of the cubic equation must therefore be doubled. This yields the coefficient representations

$$\begin{aligned} r &= rd_{2m}(2e) \\ s &= rd_{4m}((rd_{2m}e)^2 - 4rd_{4m}(g)) \\ t &= -rd_{6m}(rd_{3m}f)^2 \end{aligned}$$

of the cubic equation. The precision representation of the coefficients of the reduced cubic equation are

$$rd_{4m}(p), rd_{6m}(q) \quad and \quad rd_{6m}(D)$$

and the bottleneck precision of 10^{-12m} has to be taken for the radicand of the discriminant D, i.e., the 12-times precision of the result precision 10^{-m}.

Only possible real solutions of the arising cubic equation can produce possible real solutions of the fourth degree equation because they are only produced by sums over roots of the cubic solutions.

Summary. *The precision of geometric parameters (points of E^3) is reflected in the algebraic parameters of an equation of fourth degree with varying precision, up to 12-times precision. If programs do not use an adapted ε in domain tests or the database of a program does not make the necessary storage allocation for a sufficient large number representation, then it leads, in general, to geometrically non-suitable representations of the computer internal model.*

This summary shows only one view of correct reflections of the wanted geometric model. It is possible to improve the model by increasing the number of significant digits used in the representation. The knowledge of condition numbers allows us to discuss actual modeling systems. Since the full digit length of the hardware's arithmetic processors is to be directly addressed, linearly restricted geometric objects could have geometrically suitable representations of the desired models. A number with thirty digits is sufficient to do all necessary algebraic operation with linearly restricted geometric objects. Geometric objects of a higher algebraic degree sometimes need arithmetic for the algebraic parameters of more than a hundred digits. Complicated algebraic operations are based, as a rule, on linear or affine geometry.

A second way, for higher degree equations and without such expensive arithmetic, will be introduced later by means of a topological idea, thus modifying the concept of geometric model representations.

6.4. Quadric Intersection by Cardano's Formula

Floating point representations of surface parameters are more favorable than representations without rounding errors with regard to the increasing number of digits. Nevertheless, new questions arise by using floating-point numbers.

Should, for example, geometric objects be approximated by linearly restricted boundaries or should they be approximated by surfaces of a higher degree? Both courses can be concluded in the practice of modeling.

- Linear equations have the desired precision in the representation as well as a reserve for calculating in double precision. The large number of linear boundary parts which are necessary for the approximation is a disadvantage.

- Using geometrically suitable surfaces of a higher degree, the number of significant digits for the representation has to be increased - as yet to be motivated. Otherwise, one has to take into consideration the arising errors. Those errors can make the Boolean set operations unworkable (see Chapter 7).

In this sense, the desired increase in precision seems to be not so favorable while approximating with surfaces of a higher degree in contrast to linearly restricted approximations.

The representation of algebraic curves and points, dependent upon a geometrically suitable representation of the surfaces, deserves special attention. In general, the inaccuracy of geometric parameters increases by intersecting. Figure 6.5 shows approximation lines varying in accuracy in this pattern of the linearly approximated union of a cylinder, a frustum of a cone and a cube. All vertices are still sufficiently precise, as well as

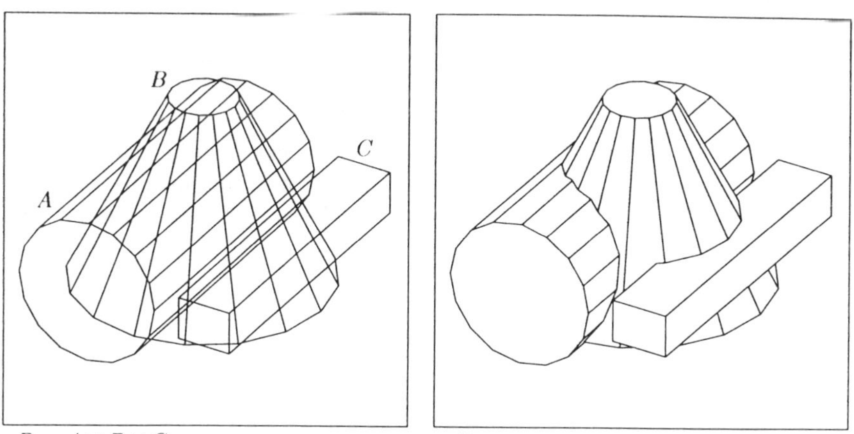

$D = A \cup B \cup C$ D

Figure 6.5. The verticies of the solids are correct nevertheless the new intersection points are not geometrically suitable.

the approximation lines, therefore it can be presumed that the surfaces are approximated geometrically suitably.

The boundary representation has non-precise inner points of the approximation lines. That can be seen at the intersection line at the cone and the cylinder.

Those problems arise in the approximation of surfaces by polynomial surfaces of second degree, as well as higher degree. But the combinatorial properties seem to be more favorable because often times essentially fewer surface pieces are needed to conclude the given definition precision. Using second degree surfaces for example, it is interesting to know which representation of intersection points has to be chosen because the significant digits of the geometric parameters (points) already double the significant digits of the algebraic parameters (equations).

A so called *quadric* is an surface of E^3 that is defined by an implicit equation of a second degree polynomial in the variables x, y, z. The intersection of two such surfaces leads to a system of equations of second degree with two equations and the three variables x, y, z. If we sort out the variable z of these equations, and divide by the coefficients of z^2, then it yields the following form of the system of equations

$$\begin{aligned} 0 &= z^2 + a_1 z + a_2 \\ 0 &= z^2 + b_1 z + b_2. \end{aligned}$$

The coefficients a_i, b_i $(i = 1, 2)$ are polynomials of degree i in the variables x and y, and we have that

$$\begin{aligned} a_1 &= a_1(x, y) = a_{11}x + a_{12}y + a_{13} \\ b_1 &= b_1(x, y) = b_{11}x + b_{12}y + b_{13} \\ a_2 &= a_2(x, y) = a_{21}x^2 + a_{22}y^2 + a_{23}xy + a_{24}x + a_{25}y + a_{26} \\ b_2 &= b_2(x, y) = b_{21}x^2 + b_{22}y^2 + b_{23}xy + b_{24}x + b_{25}y + b_{26}. \end{aligned}$$

If the division is not possible because the corresponding coefficient is zero, then we directly get the desired form with a quadratic term of the variable z only in one of the two equations. If the quadratic z-term is not vanishing in the both equations, then we transform the system of equations into

$$\begin{aligned} 0 &= z^2 + a_1 z + a_2 \\ 0 &= (b_1 - a_1)z + (b_2 - a_2) \end{aligned}$$

and eliminate z in the first equation

$$\begin{aligned} 0 &= (a_2 - b_2)^2 + a_1(b_1 - a_1)(a_2 - b_2) + a_2(b_1 - a_1)^2 \\ 0 &= (b_1 - a_1)z + (b_2 - a_2). \end{aligned}$$

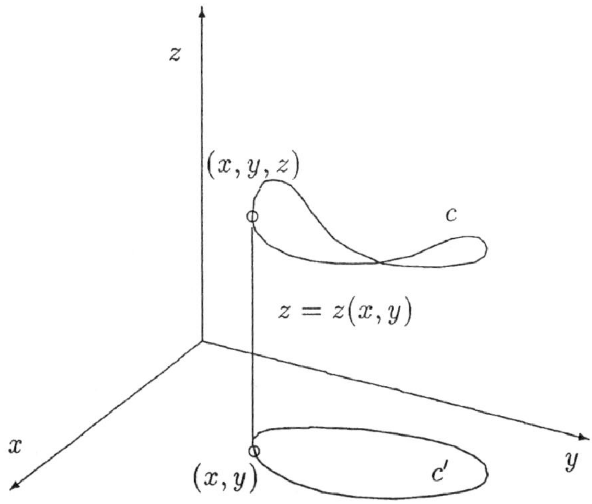

Figure 6.6. Cylindrical representation of the curve c by its projection c' and the high $z = z(x, y)$.

Now, the first equation only depends on the two variables x and y and represents the projection of the desired curve into the xy-plane. The second equation permits us to compute the length of the curve point (x, y, z) over each projection point (x, y) with $b_1(x, y) - a_1(x, y) \neq 0$, see Figure 6.6.

The problem of finding a parametric representation of the curve c can now be assumed to be the problem of finding a parametric representation of the projection curve c' in the xy-plane. We chose

$$x = x(t) = t$$

as the free parameter and find from the equation of fourth degree,

$$0 = (a_2 - b_2)^2 + a_1(b_1 - a_1)(a_2 - b_2) + a_2(b_1 - a_1)^2,$$

an implicit representation of y=y(t)

$$\begin{aligned} 0 \;=\; & (-a_{13} + b_{13} - a_{11}t + b_{11}t - a_{12}y + b_{12}y)^2 \\ & (a_{26} + a_{24}t + a_{21}t^2 + a_{25}y + a_{23}ty + a_{22}y^2) + \\ & (a_{13} + a_{11}t + a_{12}y)(-a_{13} + b_{13} - a_{11}t + b_{11}t - a_{12}y + b_{12}y) \\ & (a_{26} - b_{26} + a_{24}t - b_{24}t + a_{21}t^2 - b_{21}t^2 + \\ & a_{25}y - b_{25}y + a_{23}ty - b_{23}ty + a_{22}y^2 - b_{22}y^2) + \\ & (a_{26} - b_{26} + a_{24}t - b_{24}t + a_{21}t^2 - b_{21}t^2 + \\ & a_{25}y - b_{25}y + a_{23}ty - b_{23}ty + a_{22}y^2 - b_{22}y^2)^2 \end{aligned}$$

that we have to resolve by y. The resulting equation of fourth degree

$$0 = c_4 y^4 + c_3 y^3 + c_2 y^2 + c_1 y + c_0$$

has very large terms in the coefficients c_0, c_1, c_2, c_3, and c_4 and can be found in the appendix of this book. It was produced by analytic computations of a computer program. We can see that this equation is already so large that it seems to be undesirable to write down the whole parametric representation of the considered plane curve. We have to decide later whether or not we continue the computation with an algebraic scheme in order to give the entire parametric representation, or we approximate the curve by linear polygons such that the verticies are sufficiently precise to the curve.

The second way needs an enumeration of starting points from which we can trace the whole curve. Those starting points can be evaluated by a projection method again. We project the plane curve represented by $F(x,y) = 0$ on the x-axis and compute thereby the points whose tangents are orthogonal to the x-axis, i.e., where $F_y(x,y) = 0$. These points can be taken as starting points for tracing algorithms. The system of equations being solved,

$$\begin{aligned} 0 &= c_4 y^4 + c_3 y^3 + c_2 y^2 + c_1 y + c_0 \\ 0 &= 4c_4 y^3 + 3c_3 y^2 + 2c_2 y^1 + c_1 y, \end{aligned}$$

can easily be transformed by reciprocal substituting of the highest y-power and leads to an eigth degree equation in the variable x (or t). All of the zeros can be found by computer algebraic methods with the following idea. One computes recursively all zeros of the derivation polynomial equation. A local minimum under the x-axis and a local maximum over the x-axis defines a sufficient small interval of reliable starting point all of the inner zeros that can be iterated by methods of the Newton-type. The starting points of both possible exterior zeros can be taken apart from the lowest and highest intervals.

If we do not compute the whole parametric representation of the intersection but only a polygonal approximation by means of iteration methods, in order to trace the plane projected curve, then it is not necessary to use polynomials with a degree greater than two. The projection method to get a plane curve, and to divide the plane curve into pieces of functions can always be used only by a simple elimination of the highest power of a special variable in one of the two given equations. This method is called *cylindrical algebraic decomposition*. The practicability of the cylindrical algebraic decomposition depends strongly on correct application of the condition numbers of the algebraic parameters to get suitable solutions. That is, the ε-tests for comparison of polynomials, as well as the number of digits for rounding, have to be chosen in the sense of a parametrically continuous representation as mentioned in Chapter 7.

6.5. Tangential Quadrics as an Intersection Problem

Connecting two given curves by a special face is called *blending.* It is used for constructing a *blending face* that crosses from a given face to another one. At first we are interested in blending between two quadric surfaces.

In describing hull algebras, we often need a method of finding a surface that is defined by means of a special blending of two quadric surfaces. This surface has to be tangent to both quadric surfaces.

It is possible to define such a special surface that can be generated by straight lines. A surface is called a *ruled surface* if it contains at least a one-parameter family of straight lines that can be taken as coordinate lines of this surface. Those straight lines are called the *generating lines* . A ruled surface which has the same tangent plane in all points of one and the same generating line is called a *developable surface* . Thus, developable surfaces are those ruled surfaces which can be wound-up in a plane.

Planes, cylinders and cones are developable surfaces. Hyperbolic paraboloids and hyperboloids are only ruled surfaces. According to its definition, we can produce a ruled surface if we move a straight line continuously in the three-dimensional space. Such a motion is well-defined by a curve $x(s) \in E^3$ and a vector $z(s)$ in the direction of the straight line with the parameter s.

The Developable Surface of Two Quadric Surfaces

Let us consider now the special developable surface which is given by the motion of a plane tangent to two quadric surfaces, see Figure 6.7. We are interested in the parametric curves $x(s)$ and $y(s)$ (contact curves) with the parameter $s \in R$ with respect to each generating line s. They are well-defined by the straight line g(s) of two points $x(s)$ and $y(s)$. These curves separate the two quadric surfaces and the arising ruled surface in topologically different parts, maybe themselves non-connected parts, and the two sets of curve points $\{x(s)|s \in R\}$ and $\{y(s)|s \in R\}$. From these, new forms can be composed.

The computation of the contact curves $x(s)$ and $y(s)$ leads to the computation of an intersection of two quadric surfaces. In contrast to the description of quadric intersections mentioned above, here we will use the vector and matrix notation to derive these intersections for describing the contact curves $x(s)$ and $y(s)$.

The directions of the tangential vectors n_1 and n_2 of the two quadric areas Q_1 and Q_2

$$x^T Cx + p^T x + q = 0 \quad \textit{and} \quad y^T By + r^T y + s = 0$$

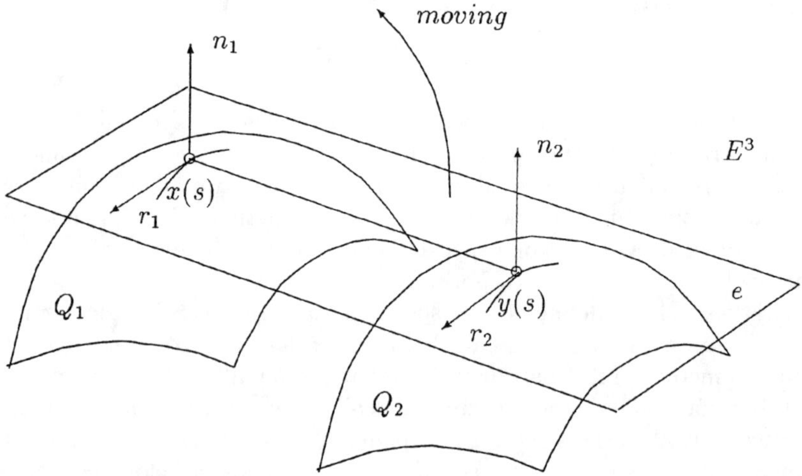

Figure 6.7. Generating a special ruled surface by two quadric surfaces.

are equal to each other, and therefore it follows with $t > 0$ that

$$\begin{aligned} (2Cx + p) &= t(2By + r) \qquad \textit{and} \\ \| 2Cx + p \| &= t \| 2By + r \| . \end{aligned}$$

The distance d of the plane e to the origin of the coordinate system can be interpreted from the quadric surface Q_1 as well as Q_2

$$d = x^T \frac{2Cx + p}{\| 2Cy + p \|} = y^T \frac{2Bx + r}{\| 2By + r \|},$$

and together we yield

$$2x^T Cx + p^T x = t(2y^T By + r^T y).$$

It follows from this equation and from the two equations defining the quadric surfaces

$$t = \frac{p^T x + 2q}{r^T y + 2s}. \tag{6.1}$$

Together we get the following description of the desired contact curves $x(s)$ and $y(s)$

$$\begin{aligned} 0 &= x^T Cx + p^T x + q & (6.2) \\ 0 &= y^T By + r^T y + s & (6.3) \\ 2Cx + p &= t(2By + r) & (6.4) \end{aligned}$$

If we put the third equation into the second one by the elimination of the variable y, then the first and the new second equation will give us the curve $x(s)$ as the intersection of two quadric surfaces. The third equation will be used to describe explicitly the curve $y(s)$ dependent on the curve $x(s)$.

The following Lemma 6.19 points out the equivalence of finding the desired contact curve $x(s)$, and it points out an intersection problem of two quadric surfaces by means of homogeneous coordinates. With this, the equivalence transformation of the two problems is easy to understand.

Lemma 6.19. *The quadric surfaces*

$$x^T Cx + p^T x + q = 0 \quad \text{and} \quad y^T By + r^T y + s = 0$$

have a common tangent plane in the points x and y if and only if the three following conditions, with homogeneous coordinates, are valid:

$$0 = \begin{pmatrix} x \\ 1 \end{pmatrix}^T \begin{pmatrix} 2C & p \\ p^T & 2q \end{pmatrix} \begin{pmatrix} x \\ 1 \end{pmatrix}, \tag{6.5}$$

$$0 = \begin{pmatrix} x \\ 1 \end{pmatrix}^T \begin{pmatrix} 2C & p \\ p^T & 2q \end{pmatrix} \begin{pmatrix} 2B & r \\ r^T & 2s \end{pmatrix}^{-1} \begin{pmatrix} 2C & p \\ p^T & 2q \end{pmatrix} \begin{pmatrix} x \\ 1 \end{pmatrix} \tag{6.6}$$

$$\text{and} \begin{pmatrix} y \\ 1 \end{pmatrix} \text{ is proportional to } \begin{pmatrix} 2B & r \\ r^T & 2s \end{pmatrix}^{-1} \begin{pmatrix} 2C & p \\ p^T & 2q \end{pmatrix} \begin{pmatrix} x \\ 1 \end{pmatrix} \tag{6.7}$$

with the factor $t = \dfrac{p^T x + 2q}{r^T y + 2s}$.

Proof. The condition (6.7) is the same as the formula (6.4). Analogously, the formula (6.6) can be derived from the formula (6.3) by means of (6.7). Then (6.5) and (6.2) are the same ones. □

It is worth noticing that the necessary inverse matrix can be computed by

$$\begin{pmatrix} 2B & r \\ r^T & 2s \end{pmatrix}^{-1} = \begin{pmatrix} (2B)^{-1} + \dfrac{(2B)^{-1} r r^T (2B)^{-1}}{2s - r^T (2B)^{-1} r} & -\dfrac{(2B)^{-1} r}{2s - r^T (2B)^{-1} r} \\ -\dfrac{r^T (2B)^{-1}}{2s - r^T (2B)^{-1} r} & \dfrac{1}{2s - r^T (2B)^{-1} r} \end{pmatrix}.$$

In this inverse matrix, only the partial inverse B^{-1} has to be known. If it is impossible to get the matrix B^{-1} then the two quadric surfaces can be exchanged. If the inverse matrix C^{-1} also does not exist, then it is

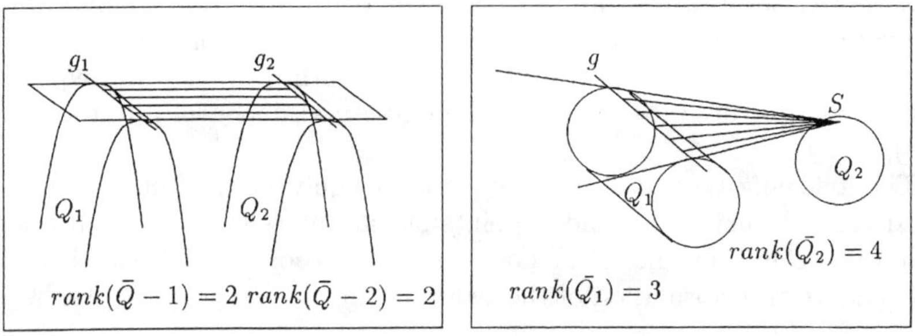

Figure 6.8. Degeneration of the blending surface by rank deficiency.

necessary to know the suitable regular part of lower dimension of the desired matrix B and to use its attached generalized inverse matrix, analogously as mentioned above.

Such cases of the quadrics and their special geometric form are indicated by the rank of the describing matrix $\bar{C}$ (or $\bar{B}$). The following three matricies, that are rank deficient, define quadric surfaces of a well-known geometrical type. They are placed in a suitable coordinate system so we can easily recognize their order.

$$\begin{pmatrix} 2 & 0 & 0 & 0 \\ 0 & 2 & 0 & 0 \\ 0 & 0 & 0 & 1 \\ 0 & 0 & 0 & 0 \end{pmatrix} \quad \begin{pmatrix} 2 & 0 & 0 & 0 \\ 0 & 0 & 0 & -1 \\ 0 & 0 & 0 & 1 \\ 0 & 0 & 0 & 0 \end{pmatrix} \quad \begin{pmatrix} 2 & 0 & 0 & 0 \\ 0 & 2 & 0 & 0 \\ 0 & 0 & 0 & 0 \\ 0 & 0 & 0 & -2 \end{pmatrix}$$

$$\begin{array}{ccc} x^2 + y^2 = z & y = x^2 & x^2 + y^2 = 1 \\ \textit{cone} & \textit{parabolic cylinder} & \textit{cylinder.} \end{array}$$

Neglecting the case that the considered quadric surfaces Q_1 and Q_2 are tangent to each other, rank deficiency leaves a blending set which belongs only to one plane or vanishes. This degeneration of the blending set can be a plane (see Figure 6.8), a straight line or only a point.

The blending definition used produces two different blending sets, which is dependent on the two given quadric equations. It is possible to multiply the equation by -1, and the two cases, shown schematically in Figure 6.9, arise.

Multiplying one of the equations by -1, the contact curves change and their orientation with respect to the orientation of the quadric surfaces changes, too.

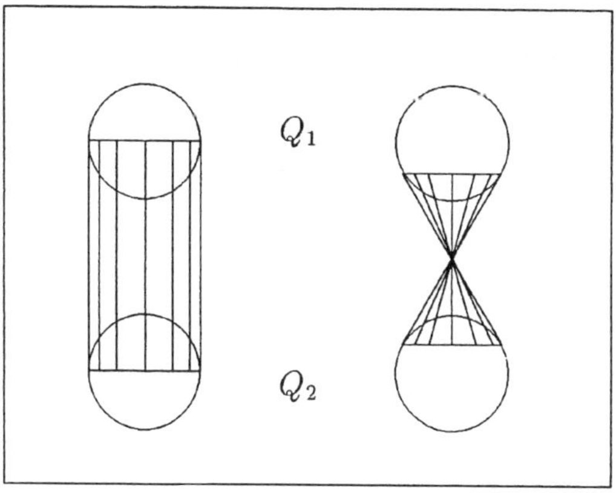

Figure 6.9. The two possible blending cases.

Ball Contact Curves of Two Quadric Surfaces

Not all contact problems with quadric surfaces can be transformed into an intersection problem. Let us consider the contact of a given ball with two quadric surfaces Q_1 and Q_2, see Figure 6.10.

The moving ball describes a so-called *ball envelope.* The midpoints of the balls define a curve which can be represented by an intersection of two *offset surfaces* of the given quadric surface Q_1 and Q_2. The offset surface of Q_1 is a surface that arises if we move all points x of Q_1 into the direction of its normal on Q_1, with a distance of the ball radius R. One can compute that

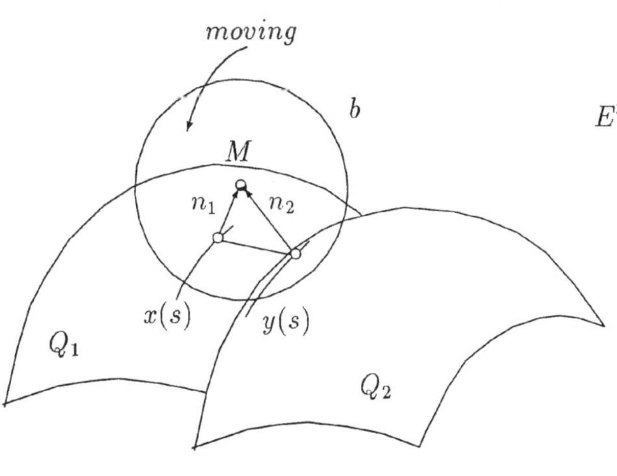

Figure 6.10. Generating of ball contact curves of two quadric surfaces.

the offset surface is of fourth degree and the plane curve after projection has therefore degree eight. Continuing the cylindrical algebraic decomposition of this intersection problem by means of polygon approximation, we get the midpoint curve and by means of translations of the polygon vertices into normal directions of the desired contact curves $x(s)$ and $y(s)$.

7

Topological Specification of Forms and Shapes

When modeling geometric objects by means of set operations, different and unwanted results occur that are dependent on the chosen representation. Unwanted results can have topological ramifications.

Set operations can be mathematically correct from the point of view of the representation, but they do not always reflect the geometric properties of an object correctly. In a typical example, improving its form yields the difference between the Boolean set operations and the regularizing set operations, or applying additional regularizing set operations as a new and independent modeling operation.

Furthermore, incorrect results can be produced or unwanted interruptions can arise in the modeling process. One reason is the inadequate stability of the modeling algorithms. Besides the instability of algorithms, model instabilities also occur (inconsistency of the model). Their causes are also of a topological nature. These model instabilities are intensified on the computer by the limited approximation of the representation of the geometric forms and shapes.

7.1. Regular Set Algebras for Representation

The result from set operations with regular polyhedra of one and the same type, needs not to necessarily be a regular one. But because of technical reasons, often only the regular part of the result of the same dimension, as the operands, are interesting. Algebras of sets with one and the same type are called *regular set algebras.*

In generating regular set algebras, we only omit in the fundamental set of a set algebra the elements of lower and higher type as wanted, and regularize the results of all finite set operations. This handling leads to a modification of the set operation. Lower dimensional parts, as the highest topological dimension of the result, are omitted and added respectively. In the algebraic sense, new difficulties occur because it seems that not all the rules of the non-regularized algebras can be transferred into the arising algebras.

A Boolean algebra does not transform into a Boolean algebra again without new assumptions. Adjustment of the addition and multiplication can be done without problems. The negation causes difficulties because the double complement

$$cl_*int_*(cl_*int_*(A'))' \neq A$$

yields another result sometimes as in the weak Boolean case ($A'' = A$).

Furthermore, if we modify the negation then the meaning of the unit elements changes, and the rules of the weak Boolean algebra are not clear. But even these rules permit the geometric interpretation of set operations by means of the Euler-Venn diagram, as a geometrically clear principle of construction and therefore they have to be used.

In principle, there are two possibilities to describe regular set algebras in a Boolean algebra $(G, \cup, \cap)$. Without neglecting aspects of application, the fundamental set G consists of a finite number of sets. Let G^o be the set of all open and G^c of all closed sets from G, and if $A \in G$ then $int_*A \in G^o$ and $cl_*int_*A \in G^c$. Then the set G^o, as well as the set G^c, can be used for representation of the sets elements of G.

Now we want to obtain algebraic closure in the set G^o and G^c by modifying the equivalence relation ($=$) of the equality of sets according to their type i $(i = 0, 1, 2, 3)$ of regularization.

Definition 7.1. *Two sets $A, B \in G$ are called* **i-regularly equal** *($i = 0, 1, 2, 3$) with respect to the*

open sets	*if and only if*	$int_iA = int_iB$	*and*
closed sets	*if and only if*	$cl_iint_iA = cl_iint_iB$,	

denoted by $A =_i^o$ and $A =_i^c B$ or only $A =_i B$.

Whether the regularization type i is larger than or equal to the type of variety of the set being regularized, the i-regular equality differs from the usual one. The principle follows from the sense of the trace topology with respect to the i-dimensional supporting set. Let, for example, A be a cubic and B its boundary then $A \neq_3 B$, $A =_2 B$, $A =_1 B$ and $A =_0 B$. If A is a rectangle and B its boundary it holds $A =_3 B, A \neq_2 B, A =_1 B$ and $A =_0 B$, and so on.

The set equality of a trace topology in a supporting set of the type i is transmitted to the whole fundamental set G:

Lemma 7.2. *The relation $(=_i)$ is an equivalence relation with respect to a supporting set of the type i in the whole set G in both cases. The relations $(=)$ and $(=_i)$ are the same ones in G^o, but with respect to the closed sets in G^c.*

Proof. Reflexivity, commutativity and transitivity with respect to open sets follow immediately from the equivalence relation $(=)$ that was used in the definition. Therefore the relation $(\overset{o}{=_i})$ is an equivalence relation to the open sets in B. Furthermore, it holds for two sets $A, B \subset B^o$ where $A = int_i A$ and $B = int_i B$, and it follows that $A = B$ if and only if $A \overset{o}{=_i} B$. These assertions also are valid with respect to the closed sets, analogously. □

It is usual in the field of computational geometry to describe geometric objects by taking regular closed sets. Topologically oriented contexts prefer open sets, the manifolds.

Let $(G, \cup, \cap)$ be the Boolean algebra mentioned above. The addition $(\cup)$ and the multiplication $(\cap)$ of two elements $A, B \in G$ are modified by a third set $Q(A, B)$ which is dependent on the both operands A and B of the operations $(\dot{\cup})$ and $(\dot{\cap})$.

Lemma 7.3. *$(G, \cup, \cap)$ is a Boolean algebra according to the new equivalence relation $(=_i)$. If the modifications $(\dot{\cup})$ and $(\dot{\cap})$ by means of the extension $Q(A, B)$ are natural ones of the Boolean algebra, then they are a natural modification with respect to the new equivalence relation $(=_i)$.*

Proof. Let $C = D$ for any two sets $C, D \in G$. Then it follows for the relatively inner points of type i $(i = 0, 1, 2, 3)$ directly that $int_i C = int_i D$, i.e., $C \overset{o}{=_i} D$. If we use the equivalence relation $(\overset{o}{=_i})$ in the axiom system of the Boolean algebra instead of $(=)$, then a Boolean algebra arises again. Furthermore it follows that the equality $(=)$ directly implies the corresponding regularization equality $(=_i)$ of the new Boolean algebra. The assertion holds, therefore, for the i-regularity with respect to the open sets. From $C = D$ also directly follows $cl_i int_i C = cl_i int_i D$ and hence, the assertion is valid with respect to the closed sets. □

The i-regular equality regarding open sets has no importance in algebras of closed sets. Even so, the i-regularity regarding closed sets has no importance in algebras of open sets. Combinations of one and the same type seem to be practical.

Definition 7.4. *Let $(G^c, \dot{\cup}, \dot{\cap})$ be a natural modeling algebra of i-regular closed sets with i-regular equality regarding the closed sets. Analogously, let $(G^o, \dot{\cup}, \dot{\cap})$ be a natural modeling algebra of i-regular open sets with i-regular equality regarding open sets.*

Objects that are to be described are not only the elements of an algebra but more generally, they are a set of elements and operations with these elements as operands, too. It is our aim to describe objects in $(G, \cup, \cap)$ by objects either from $(G^c, \dot{\cup}, \dot{\cap})$ or from $(G^o, \dot{\cup}, \dot{\cap})$; a bijective mapping ϕ from G^o to G^c has to exist that makes the operations of both algebras compatible, i.e., an isomorphism between the two algebras has to exist.

Lemma 7.5. *If $(G, \dot{\cup}, \dot{\cap})$ is a natural modeling set algebra then $(G^c, \dot{\cup}, \dot{\cap})_i$ is isomorphic to $(G^o, \dot{\cup}, \dot{\cap})_i$ for $i = 0, 1, 2, 3$.*

Proof. The mapping ϕ maps each $A \in G^o$ to $B = cl_i A \in G^c$. The map ψ^{-1} orders each closed set $B \in G^c$ exactly to the set $B = int_i B$. This mapping ϕ is injective. Obviously it holds for $A, B \in G^o$

$$\begin{aligned} cl_i(A \dot{\cup} B) &= cl_i A \dot{\cup} cl_i B \\ cl_i(A \dot{\cap} B) &= cl_i A \dot{\cap} cl_i B. \end{aligned}$$

It follows that

$$\begin{aligned} \phi(A \dot{\cup} B) &= \phi(A) \dot{\cup} \phi(B) =_i^c \phi(A \dot{\cup} B) \\ \phi(A \dot{\cap} B) &= \phi(A) \dot{\cap} \phi(B) =_i^c \phi(A \dot{\cap} B) \end{aligned}$$

with respect to the identification of the i-regular equality $(=_i^c)$ of closed sets and the equality $(=)$ in G^c. Analogously, it holds for $B, C \in G^c$

$$\begin{aligned} int_i(B \dot{\cup} C) &= int_i B \dot{\cup} int_i C \\ int_i(B \dot{\cap} C) &= int_i B \dot{\cap} int_i C. \end{aligned}$$

From this it follows

$$\begin{aligned} \phi(B \dot{\cup} C) &= \phi(B) \dot{\cup} \phi(C) =_i^o \psi(B \dot{\cup} C) \\ \phi(B \dot{\cap} C) &= \phi(B) \dot{\cap} \phi(C) =_i^o \psi(B \dot{\cap} C) \end{aligned}$$

according to the identification of the i-regular equality ($=_i^o$) of open sets and the equality ($=$) in G^o. The unit element of the multiplication (empty set) is not changed by ϕ. The unit element I^c of the addition in $(G^c, \dot{\cup}, \dot{\cap})$, as the union of all sets from G^c, is mapped by ϕ to the unit element I^o of the addition in $(G^o, \dot{\cup}, \dot{\cap})$, as the union of all sets from G^o. The complement $A' \in G^o$ from $A \in G^o$ holds because of

$$cl_i A' \;=_i^c\; (cl_i A)'$$

and

$$\phi(A') \;=\; (\phi(A))'.$$

The complement $B \in G^c$ of $B \in G^c$ holds because of

$$int_i A' \;=_i^c\; (int_i A)'$$

and

$$\psi(A') \;=_i^c\; (\psi(A))'.$$

The mappings ϕ and $\psi = \phi^{-1}$, respectively are the desired isomorphisms. □

The sets G^c and G^o can be assumed to be trivial. Because of their disjunctivity apart from the empty set, not all elements of G can be mapped one to one to G^c or G^o. The i-regularization of G arranges the map into G^c and G^o, respectively. Thereby different sets of G can be assigned to one and the same set. The compatibility between the operations of an i-algebra and the modeling algebra leads to an homomorphism.

Lemma 7.6. *$(G^c, \dot{\cup}, \dot{\cap})_i$ and $(G^o, \dot{\cup}, \dot{\cap})_i$ are homomorphic to $(G, \dot{\cup}, \dot{\cap})$ in each case $i = 0, 1, 2, 3$.*

The diagram shown in Figure 7.1 is sufficient to give a whole algebraic representation of a natural modeling set algebra $(G, \dot{\cup}, \dot{\cap})$. But it only describes the representation with respect to one specific type of regularization. Unfortunately, this description can realize only a homomorphism (only one file direction) of $(G, \dot{\cup}, \dot{\cap})$.

A general polyhedron can be represented by means of a topologically strong assembly. In this case, it is to be decided how the single regularization parts have to be connected. If the operands are of the same type, it has to be done by means of the i-regular equality of the algebras $(G^c, \dot{\cup}, \dot{\cap})$ and $(G^o, \dot{\cup}, \dot{\cap})$. This i-regular equality of the single regularization components can be compatibly continued into the whole fundamental set G.

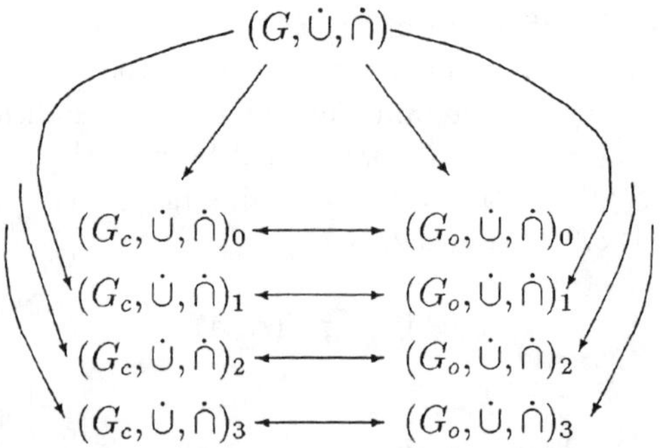

Figure 7.1. Algebraic representation of a natural modeling set algebra $(G, \dot{\cup}, \dot{\cap})$ in forms of regularized components.

Definition 7.7. *Any two sets* $A, B \in G$ *are called* regular equal $(=^c)$,$(=^o)$ *with respect to the open sets and the closed set if and only if* $A =_i^o B$ *and* $A =_i^c B$*, respectively.*

Let G be the set of all general polyhedra. Each $P \in G$ can be well-defined by a topologically strong assembly consisting of relatively closed sets. The set of all these assemblies is denoted by A^c. The set A^o contains all topologically strong assemblies of relatively open sets and each P can be well-defined analogously.

Lemma 7.8. *The algebra* $(G, \dot{\cup}, \dot{\cap})$ *of all general polyhedra with respect to the usual equality* $(=)$ *is isomorphic to the algebra* $(A^o, \dot{\cup}, \dot{\cap})$ *and* $(A^c, \dot{\cup}, \dot{\cap})$ *with respect to the regular equality* $(=^c)$ *and* $(=^o)$*, respectively.*

Proof. It holds $G = A^c$ because of the well-defined and topologically strong representation. The operations $(\dot{\cup})$ and $(\dot{\cap})$ are the same ones in the two algebras, without the different application of the equality according to their definitions. It still has to be shown that the representation $\alpha^c(A) \in A^c$ by means of regular closed parts of a set $A \in G$ realizes the equality. Let $A, B \in G$ and $\alpha^c(A), \alpha^c(B) \in A^c$. Then $A = B$ is equivalent to $\alpha^c(A) = \alpha^c(B)$ and to $\alpha^c(A)_i = \alpha^c(B)_i$ $(i = 0, \cdots, 3)$. The equality follows with respect to each trace topology, in particular the i-regular equality $(i = 0, \cdots, 3)$, too and it holds that $\alpha^c(A) =^c \alpha^c(B)$. The implication is reversible because we can assume that A and B are topologically strong assemblies. Thus $A = B$ is equivalent to $\alpha^c(A) =^c \alpha^c(B)$. The equivalence to $\alpha^o(A) =^o \alpha^o(B)$ is valid analogously. □

$$\begin{array}{ccccc}
(=) & & & & (=^c) \\
 & (G, \dot\cup, \dot\cap) & \overset{\alpha^c}{\longleftrightarrow} & (A^c, \dot\cup, \dot\cap) & \\
 & \downarrow & & \downarrow & \\
 & (G_i^c, \dot\cup, \dot\cap) & \overset{\alpha_i^c}{\longleftrightarrow} & (A_i^c, \dot\cup, \dot\cap) & \\
(=_i^c) & & & & (=)
\end{array}$$

Figure 7.2. Homomorphic transformation of the isomorphic representation map α^c into one regularization component.

The representation of general polyhedra of G by means of topologically strong assemblies, according to the isomorphic map α^c, transmit themselves by the i-regular equality in a homomorphic way into the single regularization components ($i = 0, \cdots, 3$), as shown in Figure 7.2.

The i-regular equality inside of an i-component corresponds to one and the same point equality ($=$) as in $(G, \dot\cup, \dot\cap)$. It can usually be computed with topologically strong assemblies in single regularization components. We have only collected the regular parts of the corresponding components during or after each binary operation.

Whether it can already be done during the operation in some partial results or not, depends on the validity of distributive rules that can be used in binary operation of assemblies. The distributive rules can be used unrestricted in a Boolean operation. If the extension of the modified operations is an additive and multiplicative one, then the distributions between $(\cup), (\dot\cup)$ and $(\cap), (\dot\cup)$ mentioned above can be used. Thereby, the Boolean operations are the operations that are needed for the assembly representation.

7.2. Regularization

Let us notice that the modification of a modification can be comprised into only one modification by means of the associative rule in the definition of the modified multiplication and modified addition. Furthermore, natural

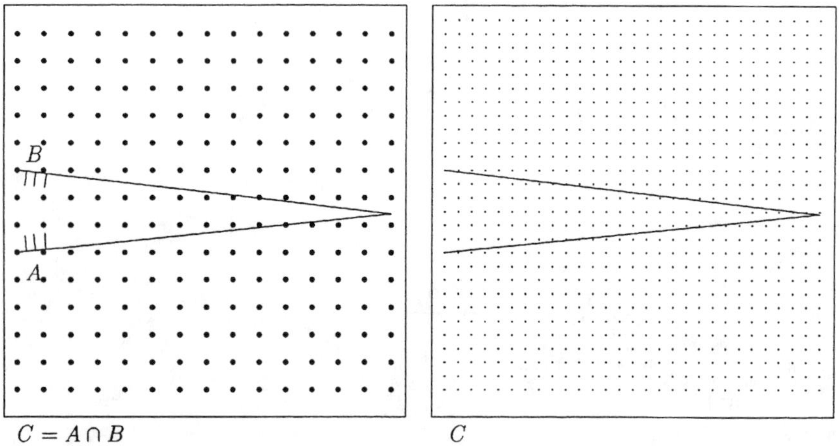

Figure 7.3. Computation of a pic by grid refinement.

modifications of binary operations can be assumed as trinary Boolean ones, that can be expressed by continued binary ones again. Boolean operations can canonically be expressed by means of basic systems like $\{(\cup), (')\}$ or $\{(\cap), (')\}$. In all of the regularization cases, the result of an operation is modified by omitting or adding parts of lower dimensions. It is only necessary in this sense to explain how a suitable i-regular (closed) polyhedron is assigned to a given general polyhedron. Let G_* be a supporting set of the type i. Then the suitable i-regular polyhedron P_i yields

$$P_i = cl_* int_* (P \cap G_*)$$

according to its definition in the sense of the i-regular equality.

It formally seems to be simple. But on a computer, especially in the case of equations of higher degree, it causes a lot of difficulties because of precision reasons. Even the question of which dimension of variety does the set $P \cap G_*$ possess may be very unclear.

If we compute a pic by increasing grid refinement, i.e., more than the representation precision EPS, it is still unclear whether the forward part of the pic seems to be a surface or should it be assumed to be a straight line (Figure 7.3). Two practical modifications are shown in Figure 7.4.

The left one describes an intersection of the pic. This modification is called *rounding regularization* of a regular polyhedron. It is not necessary to continue its special kind of natural modification because the following section describes the modification of parametric continuity, analogously.

The other modification is called *topological rounding.* The rounding regularization and the topologic rounding does not necessarily need a refinement

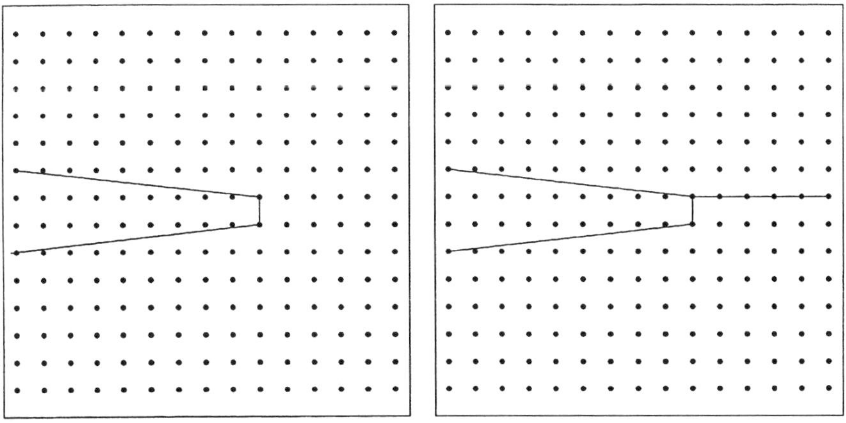

Figure 7.4. A possible regular (left) and general (right) pic modification rounded according to the representation precision EPS.

of the representation grid. We have only to accept the errors produced that are in a geometric sense smaller than the representation precision EPS.

The question whether two surfaces are equal or not has to be answered in the ε-sense, too. Then the equality of point sets has to be transmitted in some sense into the equality of algebraic parameters. The relation between geometric and algebraic parameters, according to the notion of geometrically suitable representations mentioned above (Chapter 5), is useful for testing equation parameters. If two surfaces or lines are geometrically suitable, then they are to be accepted as equal.

7.3. Parametric Continuity of Forms and Shapes

Modeling instabilities in set operations have disadvantageous consequences for the user. In order to pass over the instabilities of the model by changing the construction leads to a restriction of possibilities and it sometimes leads to a break in the automatic procedure in the modeling process.

It cannot be expected that modeling instabilities are totally eliminated. It is desirable that they influence the computing process only a little or not at all. Two demands occur hereby. Firstly, a model instability has to be recognized by the computer and secondly, each possible case has to permit a correctly readable model. The first demand can easily be fulfilled by the exact differentiation of so-called topological discontinuities. But associated incidence tests are falsified by the infinite number representation and by using ε-tests. It is required to modify the model in the sense of ε-precision.

Firstly, let us get familiar with the topological discontinuities in a special example:

Computing the intersection of two cubes A and B (see Figure 7.5) whose covers are situated in the same supporting plane. We change the coefficients in the surface equation of the covers of A and B with a sufficiently small difference δ.

With those changes by means of δ, two topologically different sets G develop for the separation of the boundaries of A and B respectively. In this case, the cover patch of the result at $\delta = 0$ runs into the straight line XY. Then the following is of strong consequences: Even if this straight line XY is determined by a stable algorithm it can diverge from the desired exact result so far that its endpoint X differs from the straight line ZX (here it is a stable one) more than the computer internal precision EPS.

Practical tests have shown that incidence tests of balls by means of usual norms (for example Euclidean norm and so on) are not sufficient to produce the corresponding relation for a correct boundary separation as well as a precondition for the new formation of the resulting boundary. In those cases additional ε-domains have to be computed and one point from the intersection of the ε-domains of both straight lines XY and ZX have to be found as the connection point X.

To imagine when those discontinuities occur and how to treat them, we extend the theoretical views of the modeling by the notion of *parameter stability.* Let us look at the closure $\bar{E}^3$ of Euclidean space E^3. This closure

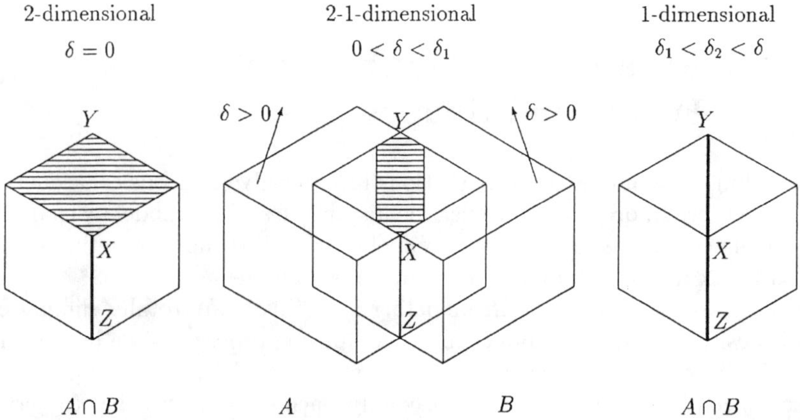

Figure 7.5. Small errors of the intersection of two cubes can lead to topological discontinuities in the result.

can be imagined by means of a stereographic projection of a sphere in the space E^4. The stereographic projection offers a homomorphic map.

Definition 7.9. *Given is a set $B \in R^m$ and the set $\mathcal{P}(\bar{E}^3)$ of all the sets of the closed three-dimensional Euclidean space. We are interested in a map f from $B \setminus \{0\}$ into $\mathcal{P}(\bar{E}^3) \setminus \emptyset$. B is called* **parameter domain**. *A special mapping set $f(u)$ with $u \in B \setminus \{0\}$ is called* **form** *of the* **shape** *f.*

The map f is a point set mapping. Such a mapping is called a *continuous set mapping* in $u \in B \setminus \{0\}$ if a $\delta > o$, a $v \in B \setminus \{0\}$ and an $y \in f(v)$ exist for all $\varepsilon > 0$ that

$$\| u - v \| \quad \textit{implies} \quad x \in U_\varepsilon(u) \quad \textit{and} \quad y \in U_\varepsilon(v).$$

The shape is called a *continuous shape* in u if there exist forms $f(v)$ of the associated shape which approach the form $f(u)$ sufficiently near in all its points.

Definition 7.10. *If the form $f(u)$ is continuous in all possible parameter vectors u then the shape is called a* **parametrically continuous shape**. *If all possible forms $f(u)$ of a shape are parametrically continuous then the shape f is called just the same.*

According to this definition no topological discontinuities occur in parametrically continuous shapes. We look at the following examples to become familiar with this notion.

Example 7.11. The linear equation

$$u_1 x_1 + u_2 x_2 + u_3 x_3 = u_4$$

defines the shape of a plane by means of all possible different definitions of its parameters $u_1, \cdots, u_4$. This shape is parametrically continuous.

Example 7.12. The intersection of any two of those special planes e_1,e_2 gives us a shape that is not parametrically continuous. A sufficient fit does exist at $e_1 = e_2$.

Example 7.13. Analogously the intersection of two half-spaces e_1^+ and e_2^+ is not parametrically continuous at $e_1^+ = -e_2^+$.

Example 7.14. Because of examples 2 and 3, the boundary representation model and the half-space model (set connections of half-spaces by means of Boolean set operations) are not parametrically continuous.

In the definition of parametrical continuity one has to pay attention to the two conditions $v \neq 0$ and $f(v) \neq \emptyset$. The condition $v \neq 0$ should give us the possibility to suspend trivial conditions in our descriptions. But that leads to a new problem. According to the definition, the shape of the following plane

$$u_1 x_1 + u_2 x_2 + u_3 x_3 = (u_4 - 1)$$

would not be parametrically continuous in $u = (0, 0, 0, 1)^T \neq 0$. If, in those cases, we substitute the parameters in such a way that the description at $v = 0$ is a trivial one, then parametrically continuous forms and shapes arise.

The second condition $f(v) \neq \emptyset$ could also be left out. It only simplifies the usage of the continuity notion of a point-set mapping and it suspends several cases which are uninteresting because of the restricted modeling space that is to be viewed.

7.4. Removal of the Model Discontinuity

The aim is now to remove the discontinuity of the forms by small modifications. Explaining the principle of these modifications, we consider firstly the problem of finding the intersection of two planes e_1 and e_2 from E^3. A discontinuity of the associated shape is given if both planes are equal to each other.

Figure 7.6 shows the two planes e_1 and e_2 as straight lines and their neighborhoods that are chosen accordingly to the object precision EPS. The straight lines $G_1, \cdots, G_4$ drawn as points are the intersections of the planes e_1 and e_2 and the boundaries of their neighborhoods.

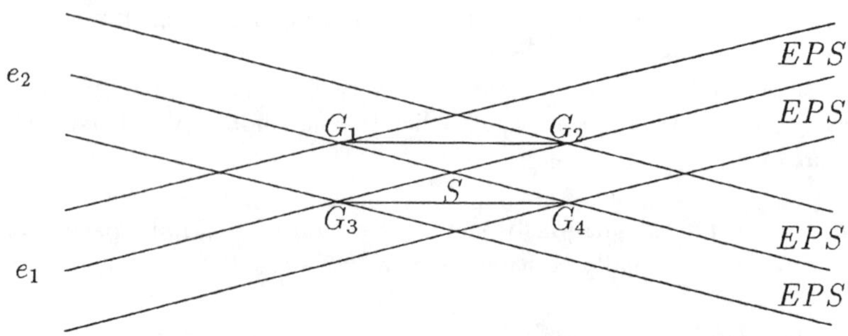

Figure 7.6. Intersection S of two planes e_1, e_2 and its possible modification as the ribbon G_1G_2 or G_3G_4.

The intersection is represented by the equations

$$\begin{array}{lrcl} e_1 & u_1x_1 + u_2x_2 + u_3x_3 & = & u_4 \\ e_2 & u_5x_1 + u_6x_6 + u_7x_7 & = & u_8. \end{array}$$

The object precision EPS is assumed to be a fixed-point number. In the sense of the ε-precision used on the computer, all points of the intersection between the two neighborhoods can be used as intersection points. A possible modification of the intersection line S is the ribbon $S_1(G_1, G_2)$ and $S_2(G_3, G_4)$ between the straight lines G_1, G_2 and G_3, G_4 respectively. It is essential that these result modifications are 2-dimensional sets like the operands e_1 and e_2. Twisting one of the planes at the straight line S causes the ribbons to increase in volume. The discontinuity of S is not a discontinuity of the ribbons $S_1(G_1, G_2)$ and $S_2(G_3, G_4)$.

The intersection and the union of two half-spaces can analogously be modified (see Figure 7.7).

The additionally applied half-space $S_1^+(e_1^+, e_2^+)$ or $S_2^+(e_1^+, e_2^+)$ is derived from the modification e_1^+ and e_2^+, respectively. The union result is extended while the intersection result is reduced.

It is successful in this sense to modify the boundary representation as well as the half-space model such that both representations are parametrically continuous in the finite domain of interest. But not all additionally used details of the modification are of interest in computations. Let us consider once more the example of Figure 7.5. It is sufficient for the computation of a correct boundary separation G of both operands, to generate only the ε-domains of the separation G. This will guarantee in each case the connections between two pieces of the separation set G. In this way we can choose any connection point of the intersection between connected ε-domains of the straight lines XY and XZ. After the separation we can forget the ε-domains.

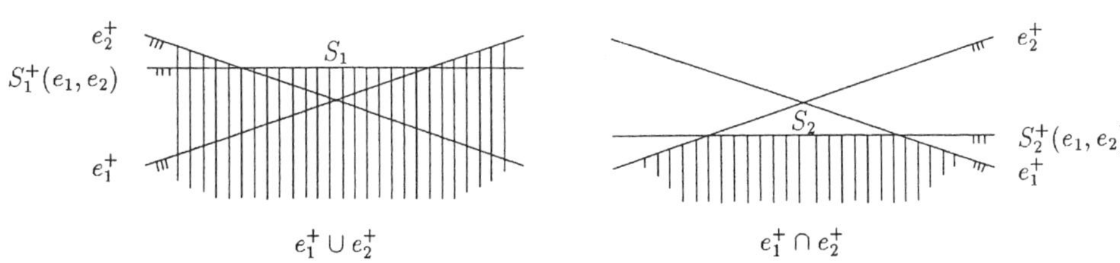

Figure 7.7. Modification of the union and the intersection of two half-spaces e_1^+ and e_2^+ by additionally using the half-spaces $S_1^+(e_1^+, e_2^+)$ and $S_2^+(e_1^+, e_2^+)$, respectively.

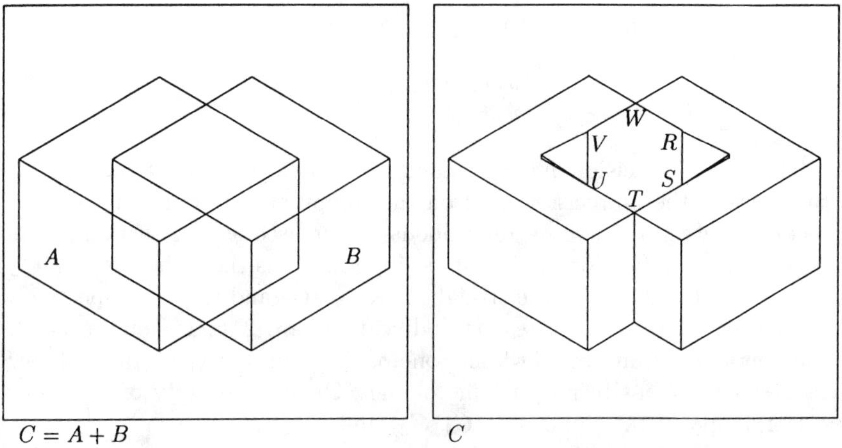

Figure 7.8. Parametrically continuous modified union of two cubes with cover patches that are not correct in the same plane.

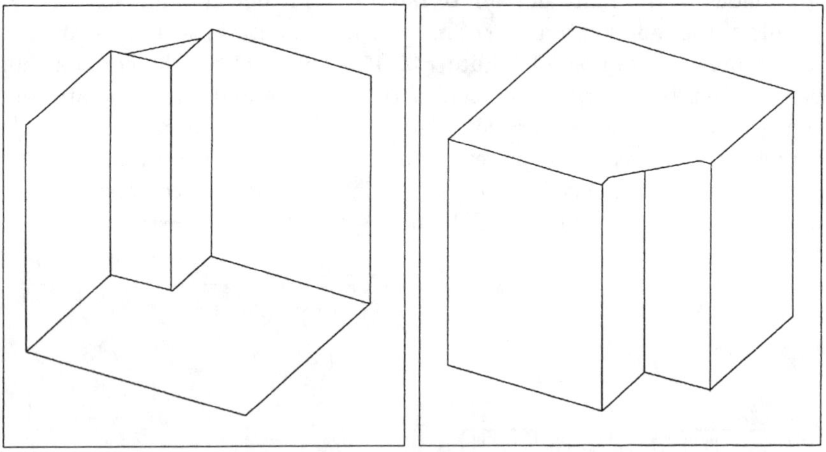

Figure 7.9. Two pictures of the difference result of two cubes with nearly the same cover patches.

Modifying the set operation in this way causes unexpected results to arise (see Figure 7.8). This is the price for reaching a correct boundary separation for an unstable model. The small overlapping pieces in the result are smaller than expected.

The example shows a whole modification of a union. If we only compute the ε-domains at the needed connections of the separation, and take for the separation a set of any connected line pieces containing the ε-domains, then a correct, but sometimes not so expected representation, is produced (see Figure 7.9).

7.5. Parametrically Continuous Set Algebras

The modifications of set operations mentioned above can be defined very differently. In its nature are all of those changes of the binary set operation ($\cup$) and intersection ($\cap$) of the operands A and B by means of a third set $Q(A, B)$. The set Q is chosen dependent on A and B and thus the modifications can be written as follows

$$\begin{aligned} A\acute{\cup}B &= (A \cup B) \cup Q(A, B) \\ A\acute{\cap}B &= (A \cap B) \cap Q'(A, B). \end{aligned}$$

Thereby the set Q' is the complement of Q. The complement operation is not modified. Altogether the Boolean algebra crosses then to the small modeling algebra. The assertions of the small modeling algebra are valid for its specification of the parametrically continuous modification, too.

It means, in the sense of parametric continuity, the validity of the following properties: The unit elements (empty set and fundamental set) of the Boolean algebra, the complement operation and also the de Morganean rule can be carried over. We lose the distributivity, absorption and idempotence. It is possible to define ε-modifications that are unsymmetrical ones, and therefore the commutativity of the union and the intersection are disturbed. Because of these cancellations, no canonical normal forms of any term, as in the Boolean algebra, can be reached. Only the weak disjunctive normal form of a binary operation, as in the small modeling algebra, can be reached. The modified union and the complement operation are enough to produce all other desired binary operation of this kind.

8

The Modeling Principle

The natural modeling algebras permit a homogeneous mathematical description for geometric modeling with respect to an algebraic and topological representation of the model. In this sense, we call this a *modeling principle*: It contains the *defining principle* of a modeling set algebra, as well as provides special *modeling elements* which are used to compute boundary representations of general polyhedra.

8.1. Modeling Set Algebras on a Computer

Until now we have stated that a natural modeling set algebra is defined by the following steps:

- looking for the Boolean set algebra $(G, \cup, \cap)$ that we have to take as the basic algebra,
- defining a natural modification $Q(A, B)$ of each binary Boolean operation of the operands $A, B \in G$ and
- building modified binary operations only by means of Boolean set operation because each modified binary operation can be assumed as a trinary Boolean one over the sets $A, B, Q(A, B)$.

This view is not sufficient to represent natural modeling algebras on a computer. It is necessary to isomorphically map the desired set algebras to logical algebras that the computer understands.

The Theorem of Stone arranges the assignment between a logic and a set algebra of Boolean type. A set $A \in G$ is mapped to the assertion $\mathcal{A} = (X \in A) \in \mathcal{G}$. According to the choice of X and with respect to A, the assertion $\mathcal{A}$ is a logical constant as $\mathcal{A} = .true.$ and $\mathcal{A} = .false.$. Then the connection of sets corresponds to the connection of logical assertions in an isomorphic sense.

The answer of the incidence question is decisive for the assignment of a logical algebra to a given set algebra. We call it *point incidence* if we want to ask whether a point belongs to a set or not. The answer to the question, whether two given sets have at least one common point or not, is called *set incidence.* Practically, it can be applied point by point if the sets have only a finite number of elements.

The Theorem of Stone can be brought into the concept of natural modeling set algebras. We understand each natural binary operation as a trinary Boolean one and we get a procedure with compatible logic. Nevertheless, the point incidence with respect to the extension $Q(A, B)$ of a binary operation $A \dot{\circ} B$ has to be clarified. The point incidence of A and B can be assumed to be known.

Defining principle. *A natural modeling set algebra can be given by defining a point incidence of the extension.*

The practicality of this principle on a computer depends on the representation used and the attached incidence tests of this representation. It should be possible to give an algorithmic answer to each incidence question.

The definition, by means of already existing modeling algebras with already implemented set representation and attached connections, is another way to describe the extension. Then the incidence question can be answered by the usual and already existing incidence algorithms.

We can assume that the extension Q is a general polyhedron. It is described by regular polyhedra of the associated topologically strong assembly. Each regular part of the extension Q is again given in principle in two ways:

- by the description of its oriented boundary, related to the associated supporting set, and

- by defining its relatively inner points by means of an incidence criterion.

In both cases we need an algorithm to decide the point incidence test of Q. The choice of the case and the desired representation lead to different algorithms for generating a computer internal representation for the extension.

8.2. Separation and Connection Sets

We want to produce modeling operations with boundary representations for a computer internal operation description of any desired natural modeling set algebra. In principle, we have to declare which parts of the operand boundaries have to be separated and which of them are to be connected, eventually by means of new sets. Furthermore, we have to decide whether the result is a correct boundary representation of a desired polyhedron or not. The goal is to describe algorithms that only produce correct boundaries and will also produce correct boundaries when an approximate interpretation of the polyhedron is used. Therefore we define the following notions of the separating set and the connection set which are already announced in the section title. Then the following two steps describe the modeling principle in the form of binary operations:

Boundary Modeling Principle

Step 1: Define the separation set and the associated half-spaces for the two operands.

Step 2: Connect, in each case, two of all the half-boundaries by means of the separation set and a special new chosen connection set depending on the separation.

Let A be an i-polyhedron $(0 \geq u \geq 3)$ and R_{ja} denote the boundary of A related to a j-dimensional supporting set $(j \geq i, 2 \geq j \geq 3)$. The boundary R_{ja} of A is separated by means of a set $S_{ja} \subset R_{ja}$

$$R_{ja} = R_{ja}^{+} \cup R_{ja}^{-} \cup S_{ja}$$

with the pairwise disjunctive sets R_{ja}^{+}, R_{ja}^{-} and S_{ja}.

Definition 8.1. *The set S_{ja} is called a* **separation set** *related to the j-dimensional supporting set if*

- *the set S_{ja} is closed with respect to a $(j-1)$-dimensional supporting set belonging to the given j-dimensional supporting set,*

- *each connected curve of R_{ja} between any $X_1 \in R^+_{ja}$ and $X_2 \in R^-_{ja}$ is not disjunctive to S_{ja} and*
- *the sets R^+_{ja} and R^-_{ja} have only a finite number of connected components.*

The sets R^+_{ja} and R^-_{ja} are called positive and negative **half-boundaries**, *respectively.*

If we define such a separation set S_{ja}, then the positive and negative half-boundaries arises. Defining only the separation sets, it does not matter which of the arising two half-boundaries of each operand becomes the positive or the negative sign. In particular, one or both half-spaces can be equal to the empty set.

For example, each closed polygon that belongs to the boundaries of a 3-polyhedron is a separation set with respect to $G = E^3$. A more difficult example of a boundary separation set of a 3-polyhedron is shown in Figure 8.1. The separation set consists of the lines $L(2, 6, 8, 4)$, $L(9, 10, 11)$ and the relatively closed fundamental set $A(1, 2, 3, 4, 1)$ of the cubic. The positive half-boundary R^+_{ja} is the relatively open part of the boundary R^-_{ja} between the two connecting parts of the separation. Then the negative half-boundaries are the relatively open parts of the other side of the separation set.

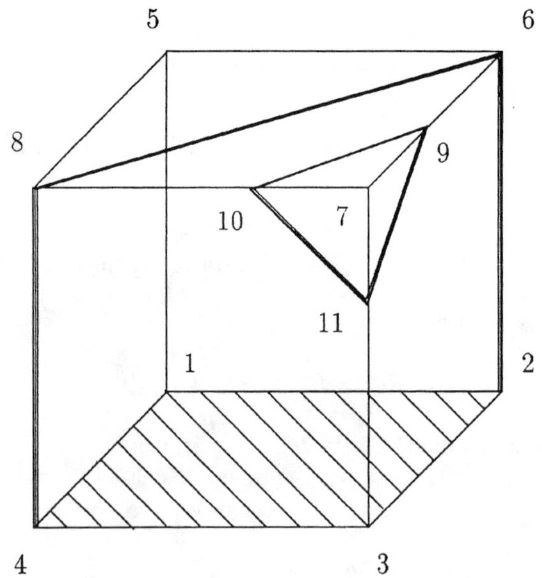

Figure 8.1. Example of a non-connected separation set of a 3-polyhedron with parts of different dimensions.

Each finite number of connected boundary point sets (for example n points) of a 2-polyhedron, related to the associated 2-dimensional supporting set, create a separation set. Another special case of a separation of a 2-polyhedron is shown in Figure 8.2.

The separation consists of the line $L(5,6)$ and the relatively closed area $A(6,7,8,9,6)$. If F_1 and F_2 are assumed as a relatively closed area then the half-boundaries according to the 3-dimensional supporting set $G = E^3$ yield

$$R_{3a}^{+} = F_1 \setminus L(5,6,9) \quad \textit{and} \quad R_{3a}^{-} = F_2 \setminus L(5,6,7,8).$$

We consider now the special kind of set which is capable of connecting half-boundaries. Let A be a simple j-polyhedron and let B be a simple k-polyhedron in a supporting set T_i. Let R_{la} and R_{mb} denote the boundaries of A and B with respect to the l-dimensional and the m-dimensional supporting set ($l, m \leq i$), respectively. S_{la} denotes a separation set of R_{la} with respect to the i-dimensional supporting set T_i, and S_{mb} denotes a separation set of R_{mb} related to the same i-dimensional supporting set T_i. Furthermore, we define an extension polyhedron $Q(A,B)$ of general polyhedron type depending on the desired operands A and B, that are simple polyhedra.

Definition 8.2. *A set C fulfills the following conditions*

$$\begin{aligned} C &= fr_i Q(A,B) \\ S_{la} \cup S_{mb} &\subset C \cup (fr_i A \cap fr_i B). \end{aligned}$$

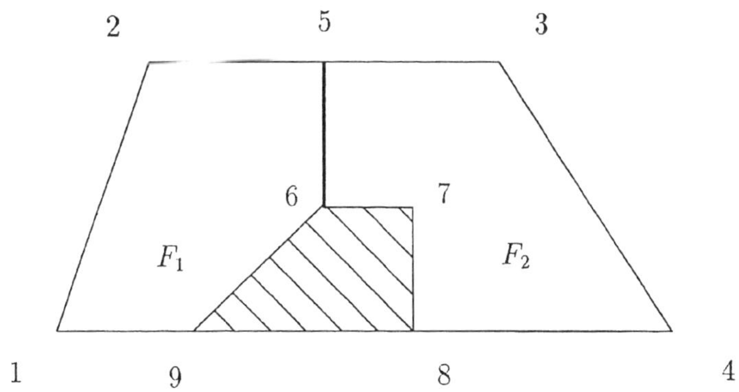

Figure 8.2. Example of a separation set of a 2-polyhedron with reference to the higher supporting set.

Then C is called a **connection boundary** *and the polyhedron $Q(A, B)$ is called a* **connection polyhedron**.

The first condition makes the relative boundary of the extension $Q(A, B)$ available to construct the relative boundary of the result of a natural modeling operation with simple operands. The second condition says that parts of the relative extension boundary can be used to continue the operand boundaries according to their separation sets.

8.3. Modeling Elements

It is our aim to describe the modeling process by means of a natural modeling algebra $(G, \dot{+}, \dot{*})$. The fundamental elements of the natural modeling algebra are general polyhedra and the set operations are modified so that we transfer the computation in such algebras to the four regularization algebras $(G_c, \dot{+}, \dot{*})_i$ mentioned above ($i = 3, 2, 1, 0$), with regular closed polyhedra and the i-regular equality. Therefore, we are interested in determining which modeling elements can describe regular computation and how the sum, carried over in terms of general polyhedra, is produced.

The dimensions representation of a general polyhedron P

$$P = P_3 \triangle P_2 \triangle P_1 \triangle P_0$$

is chosen as a useful internal representation. This representation permits computation step by step in each dimension's component. The assembly representation desired more in practice can be derived from the dimension representation.

Boolean Intersection Separation of Simple Polyhedra

We want to build the polynomial assembly representation of a simple polynomial intersection with respect to a given (oriented) supporting set. Then we have only to consider that the desired supporting set is a simple polyhedron and we want the intersection of a simple polyhedron A and a simple polyhedron B. If we describe all binary Boolean set operations by means of the basic operation system $\{\cup, '\}$, then the de Morganean rule

$$A \cap B = (A' \cup B')'$$

has to be used for describing the intersection. To build the assembly representation of polyhedra in a special supporting set the representation takes three negations and one union. Therefore it seems to be more practical to use the basic operation system $\{\cap, '\}$ to describe Boolean set operations.

Boolean Boundary Intersection of Simple Polyhedra

The simple polyhedra A of dimension j and B of dimension k are to be connected by a Boolean intersection with respect to a supporting set T_* of the dimension i that contains A and B. The set $S_i(A, B)$

$$S_i(A, B) = fr_I a \cap fr_i B$$

is a separation set of A as well as B with respect to T_*. Both of the two boundaries are decomposed into three disjunctive parts, in each case

$$\begin{aligned} fr_i A &= fr_i^+ A \cup fr_i^- A \cup S_i(A, B) \\ fr_i B &= fr_i^+ B \cup fr_i^- B \cup S_i(A, B). \end{aligned}$$

It is possible to decide whether $fr_i^+ A \subset B$ or $fr_i^+ B \cap A = \emptyset$. Without loss of generality,

$$\begin{array}{ccc} (fr_i^+ A) \cup B = \emptyset & \textit{and} & (fr_i^+ A) \cap B = \emptyset, \\ fr_i^- A \subset B & \textit{and} & fr_i^- B \subset A. \end{array}$$

We get the following Scheme 8.3 analogously to Scheme 1.2, that demonstrates a Boolean intersection of simple polyhedra. It connects the boundaries of two polyhedra of different dimensions into one and the same supporting set of higher dimension. The result does not need to be a polyhedron of a special type, i.e., the intersection produces a general polyhedron.

An assembly representation of this general polyhedron can be produced if we collect the remaining boundaries in a special kind.

- Oriented boundary parts are continued as in Scheme 8.3 by boundary parts of the same orientation.

- The remaining non oriented boundary parts yield the lower dimensional parts that can be separated by refinement of their supporting sets recursively.

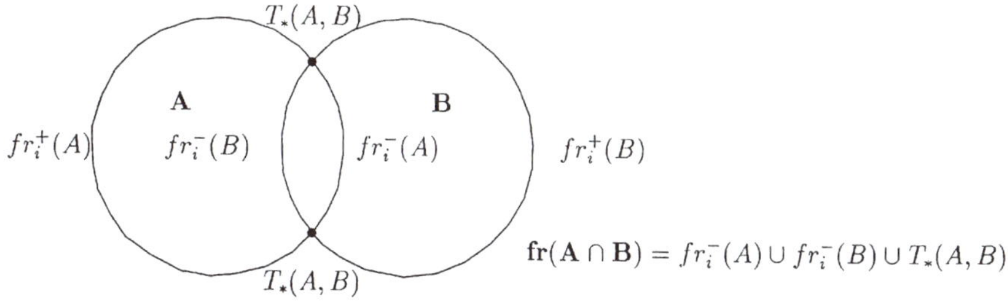

Scheme 8.3. Boolean boundary intersection of simple polyhedra of different dimensions.

The second step can be omitted if we want to do only regularized operations.

Negation of a Simple Polyhedron

If we want to make a negation of a simple polyhedron of dimension i in the natural modeling algebra $(G_c, \dot{+}, \dot{*})_i$ in the sense of the i-regular equality, then we only need to exchange the orientation of the considered simple polyhedron. In regular set algebras, the negation is very easy to realize in the computer internal descriptions.

Negation of a General Polyhedron

In the case of non-regularization, the negation result is not a simple polyhedron and we can try to negate a general polyhedron by means of regularized set operations. Therefore we ask for a negation expression of the general polyhedron P

$$P = P_3 \triangle P_2 \triangle P_1 \triangle P_0$$

with the relatively closed (simple) polyhedra P_3, P_2, P_1 and P_0 of different dimensions. We use the rules of the Boolean symmetric difference as follows:

$$\begin{aligned} P' &= (P_3 \triangle P_2 \triangle P_1 \triangle P_0)' \\ &= P_3' \triangle P_2 \triangle P_1 \triangle P_0. \end{aligned}$$

It is necessary to describe the negation of P_3 by a separation of its inner and boundary points so we can state the sum carried over into the other dimensions. The separation can be expressed again by the Boolean symmetric difference.

$$\begin{aligned} P_3' &= (P_3' \triangle fr_3 P_3) \triangle fr_3 P_3 \\ &= P_3^{*\prime} \triangle fr_3 P_3 \end{aligned}$$

with the regularized result $P_3^{*\prime}$ of the negation of P_3. The part $fr_3 P_3$ is a general 2-polyhedron and has to be carried over. Furthermore, on the general 2-polyhedron $fr_3 P_3 \triangle P_2$ is the resulting 2-dimensional remainder and

$$\begin{aligned} fr_3 P_3 \triangle P_2 &= ((fr_3 P_3 \triangle P_2)^* \triangle fr_2(fr_3 P_3 \triangle P_2)) \triangle fr_2(fr_3 P_3 \triangle P_2) \\ &= \tilde{P}_2^* \triangle fr_2(fr_3 P_3 \triangle P_2) \end{aligned}$$

whereby $\tilde{P}_2^*$ denotes the remaining (closed) regularized 2-dimensional component of P' and $fr_2(fr_3P_3 \triangle P_2)$ the 1-dimensional part that has to be carried over. The 1-dimensional component yields

$$\begin{aligned} fr_2(fr_3P_3 \triangle P_2) \triangle P_1 &= ((fr_2(fr_3P_3 \triangle P_2) \triangle P_1)^* \triangle \\ &\quad fr_1(fr_2(fr_3P_3 \triangle P_2) \triangle P_1)) \triangle \\ &\quad fr_1(fr_2(fr_3P_3 \triangle P_2) \triangle P_1) \\ &= \tilde{P}_1^* \triangle fr_1(fr_2(fr_3P_3 \triangle P_2) \triangle P_1) \end{aligned}$$

with $\tilde{P}_1^*$ as the regularized 1-dimensional component of P' and we get

$$\tilde{P}_0^* = fr_1(fr_2(fr_3P_3 \triangle P_2) \triangle P_1) \triangle P_0$$

as the remainder.

Lemma 8.3. *The general polyhedron P in dimensions representation*

$$P = P_3 \triangle P_2 \triangle P_1 \triangle P_0$$

with its relatively closed and simple polyhedra P_3, P_2, P_1 and P_0 of different dimensions has the negation P'

$$P' = \tilde{P}_3^* \triangle \tilde{P}_2^* \triangle \tilde{P}_1^* \triangle \tilde{P}_0^*$$

in dimensions representation of the same (closed) regularization type and

$$\begin{aligned} \tilde{P}_3^* &= (P_3' \triangle fr_3P_3) \\ \tilde{P}_3^* &= ((fr_3P_3 \triangle P_2)^* \triangle fr_2(fr_3P_3 \wedge P_2)) \\ \tilde{P}_3^* &= ((fr_2(fr_3P_3 \triangle P_2) \triangle P_1)^* \triangle fr_1(fr_2(fr_3P_3 \triangle P_2) \triangle P_1)) \\ \tilde{P}_3^* &= \triangle fr_1(fr_2(fr_3P_3 \wedge P_2) \triangle P_1)) \triangle P_0 \end{aligned}$$

can be computed in each case in the associated regularized Boolean set algebra.

Intersection of General Polyhedra

The intersection of simple polyhedra of dimension i in the natural modeling algebra $(G_c, \dot{+}, \dot{*})_i$ in the sense of the i-regular equality only needs the intersection of simple polyhedra according to the desired supporting set of dimension $i-1$ for describing the oriented boundary faces of the considered simple polyhedron. In regular set algebras, the intersection is often used to realize the computer internal basic system $\{\cap, '\}$.

In the case of non-regularization, the result is not a simple polyhedron and we can try to get a new general polyhedron by means of computations in each single modeling component $(G_c, \dot{+}, \dot{*})_i$ in the sense of the i-regular equality. Therefore we ask for the intersection of two expressions of the general polyhedra P and Q

$$\begin{aligned} A &= A_3 \triangle A_2 \triangle A_1 \triangle A_0 \\ B &= B_3 \triangle B_2 \triangle B_1 \triangle B_0 \end{aligned}$$

with the relatively closed (simple) polyhedra A_3, A_2, A_1, A_0 and B_3, B_2, B_1, B_0 of different dimensions. A 3-set participating share can only occur in the partial intersection of $A_3 \cap B_3$. If we regularize this partial intersection to $(A_3 \cap B_3)^*$ then we can only make a mistake on the boundary $fr(A_3 \cap B_3)$ of the partial intersection. This can be interpreted by means of the rules of the Boolean symmetric difference as follows:

$$\begin{aligned} A \cap B &= (A_3 \triangle A_2 \triangle A_1 \triangle A_0) \cap (B_3 \triangle B_2 \triangle B_1 \triangle B_0) \\ &= (A_3 \cap B_3) \triangle (A_3 \cap B_2) \triangle (A_3 \cap B_1) \triangle (A_3 \cap B_0) \\ &\triangle\ (A_2 \cap B_3) \triangle (A_2 \cap B_2) \triangle (A_2 \cap B_1) \triangle (A_2 \cap B_0) \\ &\triangle\ (A_1 \cap B_3) \triangle (A_1 \cap B_2) \triangle (A_1 \cap B_1) \triangle (A_1 \cap B_0) \\ &\triangle\ (A_0 \cap B_3) \triangle (A_0 \cap B_2) \triangle (A_0 \cap B_1) \triangle (A_0 \cap B_0) \\ &= \tilde{P}_3 \triangle \tilde{P}_2 \triangle \tilde{P}_1 \triangle \tilde{P}_0. \end{aligned}$$

Thereby, the polyhedra $\tilde{P}_3, \tilde{P}_2, \tilde{P}_1$ and $\tilde{P}_0$ are introduced as

$$\begin{aligned} \tilde{P}_3 &= (A_3 \cap B_3) \\ \tilde{P}_2 &= (A_3 \cap B_2) \triangle (A_2 \cap B_3) \triangle (A_2 \cap B_2) \\ \tilde{P}_1 &= (A_3 \cap B_1) \triangle (A_1 \cap B_3) \triangle (A_2 \cap B_1) \triangle (A_1 \cap B_2) \triangle (A_1 \cap B_1) \\ \tilde{P}_0 &= (A_3 \cap B_0) \triangle (A_0 \cap B_3) \triangle (A_2 \cap B_0) \triangle (A_0 \cap B_2) \triangle (A_1 \cap B_0) \triangle \\ &\quad (A_0 \cap B_1) \triangle (A_0 \cap B_0). \end{aligned}$$

These polyhedra are not simple ones in general, although the operand polyhedra and the intersection polyhedra are simple ones. Its lower dimensioned parts have to be carried over into the lower dimensions.

It is necessary and also possible to describe these polyhedra $\tilde{P}_3, \tilde{P}_2, \tilde{P}_1$ and $\tilde{P}_0$ by a separation of their inner and boundary points in each case. The neutral terms $fr_i\hat{P}_i \triangle fr_i\hat{P}_i$ $(i = 3, 2, 1)$ are added so that one of these two terms will explain the sum carried over and the other gives the competition of $\hat{P}_i$ to its associated i-regular result $\hat{P}_i^*$

$$\begin{aligned}
\hat{P}_3 &= (\tilde{P}_3)^* \\
\hat{P}_2 &= (\tilde{P}_2 \triangle fr_3 \tilde{P}_3)^* \\
\hat{P}_1 &= (\tilde{P}_1 \triangle fr_2(\tilde{P}_2 \triangle fr_3 \tilde{P}_3))^* \\
\hat{P}_0 &= (\tilde{P}_0 \triangle fr_1(\tilde{P}_1 \triangle fr_2(\tilde{P}_2 \triangle fr_3 \tilde{P}_3)))
\end{aligned}$$

and we get the new description of $A \cap B$

$$A \cap B = \hat{P}_3 \triangle \hat{P}_2 \triangle \hat{P}_1 \triangle \hat{P}_0$$

by means of the symmetric differences.

Lemma 8.4. *The general polyhedra A and B in dimensions representation*

$$\begin{aligned}
A &= A_3 \triangle A_2 \triangle A_1 \triangle A_0 \\
B &= B_3 \triangle B_2 \triangle B_1 \triangle B_0
\end{aligned}$$

with their relatively closed and simple polyhedra A_3, A_2, A_1 and A_0 B_3, B_2, B_1 and B_0 of different dimensions have the intersection $A \cap B$

$$A \cap B = \hat{P}_3 \triangle \hat{P}_2 \triangle \hat{P}_1 \triangle \hat{P}_0$$

in dimensions representation of the same (closed) regularization type and

$$\begin{aligned}
\hat{P}_3 &= (\tilde{P}_3)^* \\
\hat{P}_2 &= (\tilde{P}_2 \triangle fr_3 \hat{P}_3)^* \\
\hat{P}_1 &= (P_1 \triangle fr_2(\hat{P}_2)^* \\
\hat{P}_0 &= (\tilde{P}_0 \triangle fr_1(\hat{P}_1).
\end{aligned}$$

Each of the polyhedra $\hat{P}_i$ $(i = 3, 2, 1, 0)$ can be computed in the regular (closed) Boolean set algebras $(G_c, +, *)$ again

$$\begin{aligned}
\tilde{P}_3 &= (A_3 \cap B_3) \\
\tilde{P}_2 &= (A_3 \cap B_2) \triangle (A_2 \cap B_3) \triangle (A_2 \cap B_2) \\
\tilde{P}_1 &= (A_3 \cap B_1) \triangle (A_1 \cap B_3) \triangle (A_2 \cap B_1) \triangle (A_1 \cap B_2) \triangle (A_1 \cap B_1) \\
\tilde{P}_0 &= (A_3 \cap B_0) \triangle (A_0 \cap B_3) \triangle (A_2 \cap B_0) \triangle (A_0 \cap B_2) \triangle (A_1 \cap B_0) \triangle \\
&\quad (A_0 \cap B_1) \triangle (A_0 \cap B_0).
\end{aligned}$$

Boundary of an Extension $Q(A, B)$ of Polyhedra A and B

The extension polyhedron $Q(A, B)$ can be defined very differently and therefore we have to assume that it is of general polyhedron type. Sometimes it is possible to transmit the computation of the extension $Q(A, B)$ into terms of regularized operation. It takes place if the extension $Q(A, B)$ is additive and multiplicative, and because of

$$Q(U,V) = Q(U',V) = Q(U,V') = Q(V',U')$$

for all U and V according to the definition of the extension $Q(U,V)$. Namely, it follows that

$$\begin{aligned} Q(U\triangle V, W) &= Q((U'\cap V)\cup(U\cap V'), W) \\ &= (Q(U',W)\cap Q(V,W))\cup(Q(U,W')\cap Q(V',W)) \\ &= Q(U,W)\cap Q(V,W) \end{aligned}$$

and together it yields the following.

Lemma 8.5. *If the extension $Q(A, B)$ of two polyhedra A and B*

$$\begin{aligned} A &= A_3\triangle A_2\triangle A_1\triangle A_0 \\ B &= B_3\triangle B_2\triangle B_1\triangle B_0 \end{aligned}$$

in dimensions representation is additive and multiplicative then

$$Q(A,B) = \bigcap_{i,j} Q(A_i, B_j).$$

Of course, the assertion about additive and multiplicative extensions $Q(A, B)$ are obvious but they seldom occur in practice. Therefore, as a rule, any monotone operation of a modeling algebra is made by means of the Boolean operation. Scheme 8.4 shows the pairwise separation set of the three polyhedra of each monotone operation.

8.4. Principle and Specification

The *modeling principle* consists of the following ideas:

- The dimensions representation of polyhedra contains the separation of different dimension parts of a polyhedron.
- Boolean computation with regular sets in a regular algebra $(G, +, *)_i$ of dimension type i can be expressed by the regularizing intersection and negation of simple polyhedra. The i-regular equality of the

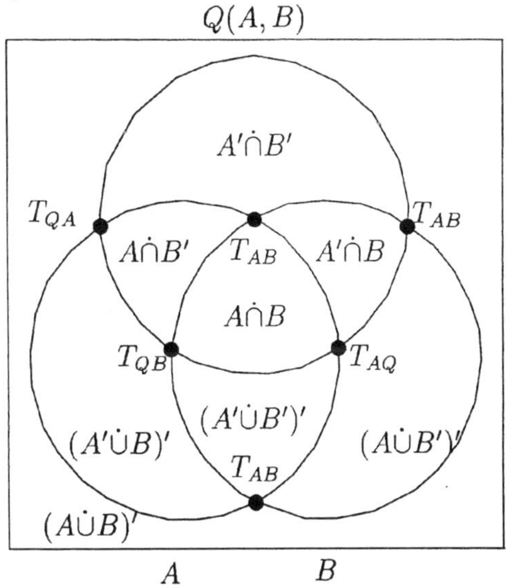

Scheme 8.4. Boundary separation of a naturally modified intersection of simple and different dimensional polyhedra.

desired algebra can be expressed by regularizing negation without neglecting our understanding of a negation in the axiomatic sense. Our aim is to achieve the same constructive meaning of the complements of a set being modeled.

- Boolean computation in the algebra $(G, +, *)$ with general polyhedra is interpreted as connected i-regular computations with simple polyhedra only by means of a set formula connection.

- Computation in a natural modeling algebra $(G, \dot{+}, \dot{*})$ can be understood as a Boolean problem if we know the desired representation of the specific extension set Q.

- Natural modifications of natural modifications can be expressed by natural modifications again.

A *specification* of the principle defines the concrete algebra in which we want to model. The entire modification has to be built in generality from the following consecutive modifications, starting from a well-defined Boolean algebra $(G, +, *)$:

- The embedding of Euclidean space E^3 into its discretization leads to a natural modification.

- The internally used algebraic parameters have to be adapted to geometric parameters in the ε-sense of geometrically suitable representations, i.e., in the form of a natural modification.
- Dimensionally differing parts of each representation have to be selected and stored in the sense of i-regularization by means of a special natural modification.
- If the computational precision is not sufficient to remove model errors of parametric discontinuity then we have to modify the model into a better one by means of a natural modification.
- Only now, we can consider new constructive (natural) modifications of the Boolean algebra, for example, hull algebras, visualization algebras and so on.

8.5. Some Constructive Modeling Algebras

Boolean Algebras

The relationship between the Boolean algebra $(B, +, *)$ and the natural modeling algebra $(B, \dot{+}, \dot{*})$ has been discussed. This relationship is also valid for the set algebras $(\mathcal{B}, \cup, \cap)$ and $(\mathcal{B}, \dot{\cup}, \dot{\cap})$. In particular, the modeling algebra $(\mathcal{B}, \dot{\cup}, \dot{\cap})$ yields a special Boolean algebra $(\mathcal{B}, \cup, \cap)$ if the extension $Q(C, D)$ vanishes for all $C, D \in \mathcal{B}$. The modified union $(\dot{\cup})$ is transferred into the Boolean union $(\cup)$ thereby, and the modified intersection $(\dot{\cap})$ always has a vanishing result. The Boolean intersection $(\cap)$ can be found as a special case of the modified, binary operations of the modeling algebra.

Considering set algebras in E^3, topological properties appear in the algebraic ones. These topological properties motivate different Boolean set algebras.

The Boolean algebra $(\mathcal{B}, \cup, \cap)$ is mapped to the *i-regularized Boolean set algebra* $(\mathcal{B}_\rangle, \cup_i, \cap_i)$ if each set $A \in \mathcal{B}$ is intersected with the union D of a finite number of given supporting sets of one and the same dimension i and if the Boolean operations are exchanged by corresponding regularized operations. The so called *solid modelers* use such a 3-regularize algebra for constructing geometric objects.

The *regularized Boolean algebra* is a more general one. It operates with i-regularized objects of one and the same dimension, as well as different dimensions, and regularizes in each case the entire result according to the highest, non-vanishing dimension of the resulting parts. Modelers of this kind are recognized by a test of the union and the intersection; for an example of a cylinder A with a plane B, see Figure 8.5.

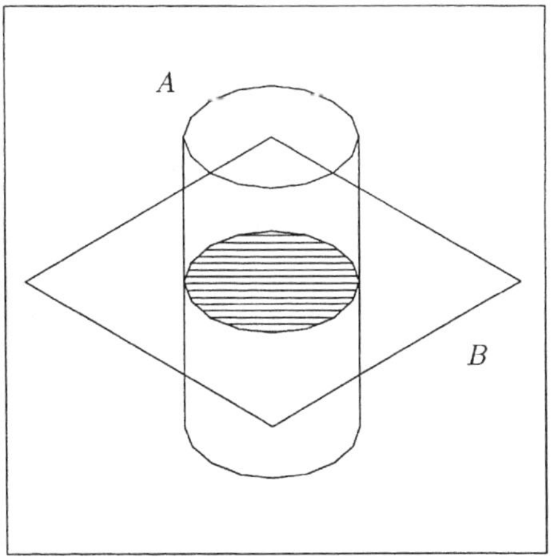

Figure 8.5. Regularization test of union and intersection.

If the result of the intersection is equal to the Boolean intersection $A \cap B$, then the operation cannot be based on a *i*-regularized algebra. If the result of the union is equal to the cylinder itself then the operation cannot be based on an algebra without regularization. Modelers with a regular Boolean algebra have to distinguish between polyhedra of different types in its data structures. This can be a useful feature for the user.

More desirable is the use of a *Boolean algebra with general polyhedra.* The general polyhedra can be represented by assemblies. Such a general modeler has therefore, as a necessary condition, to operate with assemblies of *i*-regularized parts. It has to use, at least for the internal connection of the dimensionally different assembly parts, the regularized Boolean algebra, which again internally uses a *i*-regularized Boolean algebra. In implementing a general Boolean set algebra, it is advantageous to offer the user, without any additional trouble, the *i*-regularized, as well as only the regularized Boolean algebra.

The Convex Hull Algebra

The computation of convex hulls over one or more geometric objects is practiced often. Such an operation can be done for several objects, during the binary operations. Some of the operands may be empty or of different dimensions. Therefore, the modelers with convex hull operations use such a data structure (internally) that can work with assemblies.

In the case of linearly restricted polyhedra, only the polyhedron vertices have to be used for constructing the resulting boundary representation. The

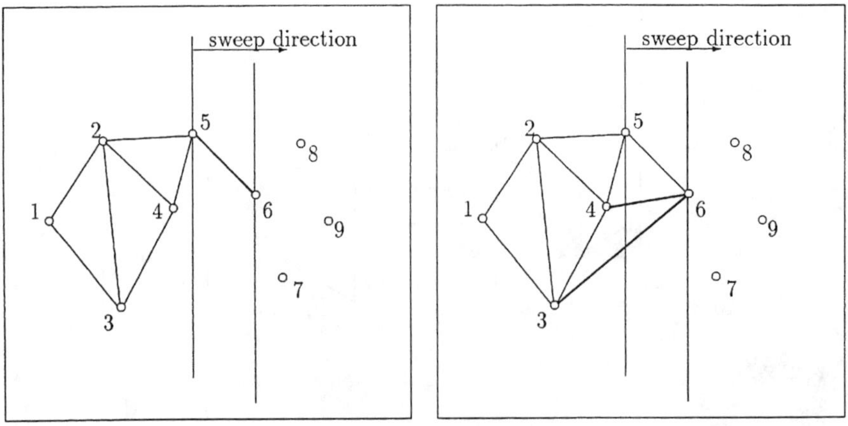

Figure 8.6. Forward step and backtracks to compute the convex hull by triangulation.

lowest possible time complexity could be explained by means of a sweep plane algorithm. Any sweep plane, chosen by a sweep direction vector, defines an ordering of the points of the considered space E^3. By adding a point, the set of points which have to be enveloped can be triangulated or formed into tetrahedrons. The *forward step* defines thereby the connection between the new point and the partial triangulation and the *backtracks*, which adds the new triangle (or tetrahedron) (see Figure 8.6).

The number of forward steps is not larger than the number n of the points to be enveloped. The total number of backtracks until the triangulation (subdivided into tetrahedra) cannot overflow the total number of triangle or tetrahedron edges. This number of vertices, edges and areas can be estimated by means of the Eulerean formula, and therefore we get in such graphs only $O(n)$ vertices, edges and surfaces. The total number of backtracks is $O(n)$.

If we delete the inner vertices (or surfaces) with $O(n)$ operations, then the desired convex hull arises. Assuming that we need $O(nlogn)$ operations to produce the order of the points by means of the sweep plane, then the total complexity of the convex hull computation is the same as of the ordering computation.

Algorithms to give the convex hull of objects with a boundary of higher degree are more difficult. They have to take not only the vertices as input data, but additionally the edges and contour lines of each boundary piece.

The Connection Hull Algebra

Let us consider $\bar{E}^3$ as the closure of E^3 by the element ∞. A set $A \in \bar{E}^3$ and the attached complement A' decomposes the fundamental set $\bar{E}^3$ such that either $\infty \in A$ or $\infty \in A'$. For practical reason, we are interested in polyhedra in A with a finitely restricted boundary and define the *finite polyhedron part*

$$finite(A) = \{ \begin{array}{ll} A & if \quad \infty \in A \; or \; A = \bar{E}^3 \\ A' & otherwise \end{array}$$

of a set A according to a decomposition of the fundamental set $\bar{E}^3$.

We introduce the following *connection hull algebra* by defining the modified union $\dot{\cup}$. Let, for any two polyhedra A, B, the modified union $A \dot{\cup} B$ be

$$A \dot{\cup} B = \bigcup_{y_1, y_2} \overline{y_1 y_2}$$

with $y_1 \in finite(A)$ and $y_1 \in finite(B)$. This is the Boolean union of all straight lines between $y_1 \in A$ and $y_2 \in B$. The extension $Q(A, B)$ of the modified algebra yields

$$Q(A, B) = (A \dot{\cup} B) \cap (finite(A) \cup finite(B))$$

and

$$A \dot{\cup} B = A \cup B \cup Q(A, B)$$
$$A \dot{\cap} B = A \cap B \cap Q(A, B)$$

are the defining operations of a modeling algebra.

The finite part $finite(A)$ of a set A was chosen therefore to obtain that

$$Q = Q(A, B) = Q(A', B) = Q(A, B') = Q(A', B') ,$$

see Figure 8.7.

Lemma 8.6. *The algebra* $(\mathcal{B}, \dot{\cup}, \dot{\cap})$ *is a natural modeling algebra.*

Proof. We have to show that the modified operations $(\dot{\cup})$ and $(\dot{\cap})$ are natural ones. If $A, B, Z \in \{\emptyset, V\}$, where V is the union over all $A \in \mathcal{B}$, and because $finite(A) = finite(A')$ it follows that

$$Q(A, Z) = Q(A', Z) \qquad Q(B, Z) = Q(B', Z) .$$

Without loss of generality, it can be assumed that

$$Q(X, Y) = \emptyset \;\; if \;\; X \notin \{A, A', V\} \;\; or \;\; Y \notin \{B, B', V\} .$$

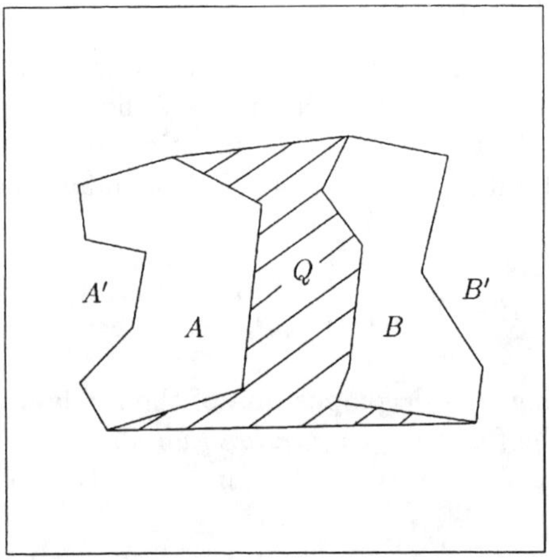

Figure 8.7. Changing the boundary orientation of operands leads to the same extension Q.

Furthermore, it holds according to the definition of the extension Q that

$$\begin{aligned} Q(\emptyset, V) &= Q(V, \emptyset) = \emptyset \\ Q(\emptyset, Z) &= Q(Z, \emptyset) = Q(\emptyset, \emptyset) = \emptyset \\ Q(V, Z) &= V \cap (finite(V) \cup finite(Z)) \\ &= Q(Z, V) = Q(V, V) = V \ , \end{aligned}$$

i.e., the modifications are natural ones. □

A special idea to compute the extension $Q(A, B)$ and the modified union $(A \dot{\cup} B)$ is the following one. From the validity of the additivity of the extension $Q(A, B)$,then

$$Q(A_1 \cup A_2, B) = Q(A_1, B) \cup Q(A_2, B)$$

follows the distribution

$$(A_1 \cup A_2) \dot{\cup} B = (A_1 \dot{\cup} B) \cup (A_2 \dot{\cup} B)$$

of the operations $(\cup)$ and $(\dot{\cup})$. Now, if a decomposition of the polyhedra or their surfaces are known, then a convex hull algorithm together with a union algorithm of the Boolean type are able to produce the desired extension. The Figure 8.8 shows objects that are constructed in this way.

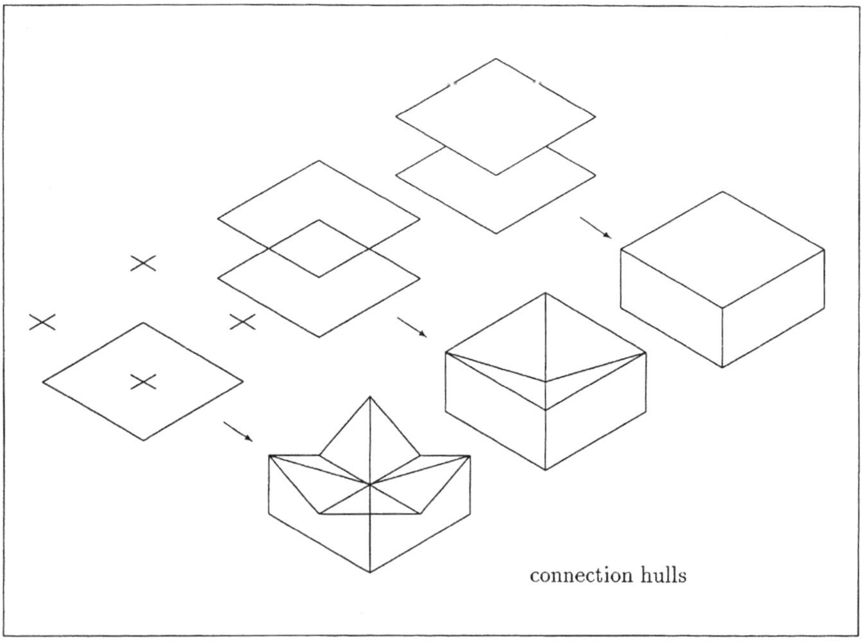

Figure 8.8. Connection hulls with dimensionally different operands.

Another way to compute the extension $Q(A, B)$ and the modified union $(A\dot{\cup}B)$ uses the construction of the connection set to represent the boundary of the extension. Those algorithms can be simplified if we modify the definition of the connection hull from above.

Modified Connection Hull Algebras

There are many possibilities to modify the connection between the two operands A and B (see Figure 8.9).

- The *visualization hull* connects visible points between the two operands.
- The *shortest connection hull* defines that part of the connection hull which connects the two operands.
- The *left distance hull* connects the points of the first operand A with points of the second operand B with a shortest distance.
- The *right distance hull* is a left one with exchanged operands.
- The *distance hull* is the union of the left and right one.
- The *rounding hull* is defined by means of round objects that touch both operands.
- The *separation (or union) hull* is a ribbon with maximal (or minimal) width which separates the two operands.

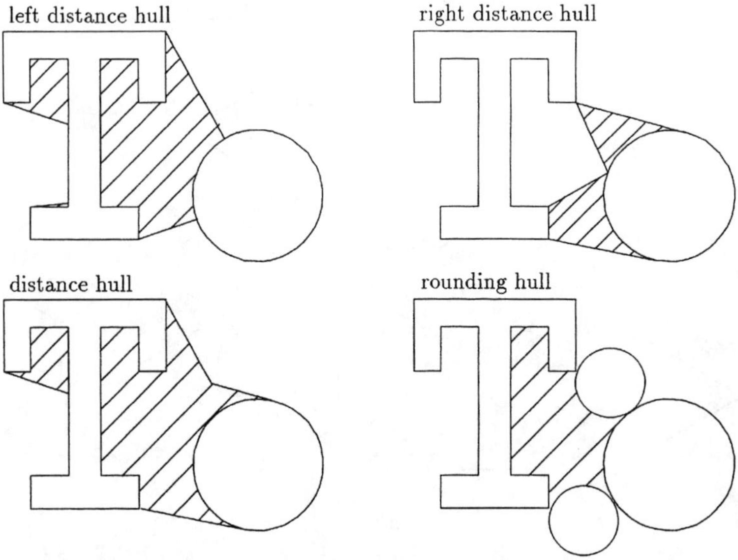

Figure 8.9. Some possible modified hull definitions which may be used to define special constructive modeling algebras.

The process of constructing additional hulls can be continued. Each of these hulls may be used to define a special modeling algebra.

A

The GECO Language

The geometric language that we write within a dialog, as well as into a language file, contains instructions of different importance:

- computation with numbers and vectors,
- generation of geometric objects,
- object manipulation,
- controls, input and output.

The Dialog, Help and the Language Files

The dialog understands all commands of *GECO*. A special command permits language files that have been prepared with an editor to be processed; the processing proceeds top down, beginning with a special named language row.

$$
\begin{array}{lcl}
help & ::= & HELP \\
file & ::= & ROW \;\; integer\_1 \;\; name \;\; FROM \;\; integer\_2 \\
name & ::= & letter < letter \;|\; digit >< 0:3 >
\end{array}
$$

$$
\begin{array}{lcl}
dialog & ::= & \left\{ \begin{array}{l} dialog\_begin \\ dialog\_end \\ help \\ file \end{array} \right\} \\
dialog\_begin & ::= & DIALOG \\
dialog\_end & ::= & END\ DIALOG
\end{array}
$$

Semantics:

1. HELP calls a menu to show the details of the language description.

2. Files start with the *file* instruction. All rows that contain a double cross in the first column are accepted. A blank in the first column signifies a comment. Each command is finished by a ';', and after that it is possible to write a comment. The processing of a language file is finished when '/' is detected in the first column.

3. The row number is given by *integer_1*. The parameter *name* defines the name of the desired file with the type $'.Tn'$ whereby n defines the catalogue number corresponding to the logical device number *integer_2*. The following example calls a file $MOP1.T01$, starting with its first row.

4. Names consist of at most four letters.

5. END DIALOG can be used only during the dialog. If you finish the work with a break $< CRL \;\; C >$, then the modeling storage from the beginning of the work is used in the next session. If you end with the command END DIALOG, then the internal modeling storage in the next session is the same as in the last session.

Example:

$$
\begin{array}{l}
HELP; \\
ROW \;\; 1 \;\; MOP1 \;\; FROM \;\; 1; \\
DIALOG; \\
END \;\; DIALOG;
\end{array}
$$

Number Arithmetic

$$declaration \ ::= \ < \ INT \ \mid \ REAL \ > vardef\_list$$

$$\begin{array}{lll} vardef & ::= & var < = \ number \ >< 0:1 > \\ var & ::= & < \ name \ \mid \ ind\_var \ > \\ ind\_var & ::= & name \ [\ < integer \ \mid \ name > \ ] \end{array}$$

Remarks:

1. Arithmetic variables are from the type INT (integer) or $REAL$ (real). If they were not declared before the first application then they are declared implicitly as $REAL$. Each variable may be declared only once.

2. A variable can be allocated an initial value by *number*. The default is 0. or 0 .

3. In declarations of indicated variables, only constants as indices are possible.

Example:

$$\begin{array}{rcl} INT \ \ I & = & 1, J, K; \\ REAL \ \ P, Q & = & 2.1, R\,[\,5\,]; \end{array}$$

The Assignment Statement and the Arithmetic Expression

$$assignment \ ::= \ arit\_var \ = \ term$$

$$\begin{aligned}
arit_var \ &::= var < \ . \ name \ | \ . \ integer >< 0:1 > \\
term \ &::= \begin{Bmatrix} + \\ - \end{Bmatrix} < 0:1 > elem_term < \begin{Bmatrix} + \\ - \\ * \\ / \end{Bmatrix} elem_term >< 0: > \\
elem_term \ &::= (\ term \ ) \ | \ arit_var \ | \\
& \quad number \ | stand_fct \ (\ var \ ) \\
number \ &::= integer < \ . \ integer >< 0:1 > \\
& \quad exponent < 0:1 > \ D < 0:1 > \\
exponent \ &::= E < + \ | \ - >< 0:1 > integer \\
integer \ &::= digit < 1:6 > \\
stand_fct \ &::= SIN \ | \ COS \ | \ TAN \ | \ COT \ | \ ASIN \ | \ ACOS \ | \\
& \quad ATAN \ | \ ACOT \ | \ EXP \ | \ LOG \ | \ SQRT \ | \ ABS
\end{aligned}$$

Semantics:

1. The assignment statement assigns a value to an integer or real variable.
2. The standard priority rules are used in operations with arithmetic expressions.
3. The i-th component of a point variable X is denoted by $X.i$.
4. The symbol D can be written behind an angle (without a blank) to switch the interpretation of the angle size to degrees.
5. In describing geometric parameters, the precision and the domain of the real and integer numbers are the same as in F77.

Examples:

$$\begin{aligned}
A &= 2.1E-5; \\
FG &= X/(X+Z); \\
I &= I+1; \\
A[\ I\ ] &= SQRT(X); \\
P.1 &= Q.3; \\
C &= P.2;
\end{aligned}$$

Precision of Geometric Objects

$$detail \quad ::= \quad DETAIL \quad : \quad term$$

Semantics:

DETAIL defines the number of straightlines that are used for the linear approximation of a circle. The other non-linear objects are approximated automatically by linear restrictions too, depending on this approximation parameter. The standard value of *term* is 16. The drawing precision also corresponds to the this parameter. It can appear independently of the modeling precision of the circle division. In this way, we can differentiate between approximation lines or constructive lines while drawing. Geometric objects are stored with coordinates from -1000 until 1000. It uses a mantissa of only six decimals. The smallest absolute value 0.01.

Examples: $DETAIL: \quad N + 5 * A;$

Points, Straightlines and Planes

$$def\_point \quad ::= \quad DEF \quad var\_1 \quad = \quad point$$
$$\left\{ \begin{array}{l} (\ term \ , \ term < \ , \ term >< 0:1 > \ ) \\ INTERSECT \ var\_2 \ , \ var\_3 \ term \ FROM \ var\_2 \end{array} \right\}$$

$$def\_straight \quad ::= \quad DEF \quad var\_1 \quad = \quad STRAIGHT$$
$$\left\{ \begin{array}{l} point\_1 \ , \ point\_2 \\ INTERSECT \ var\_2 \ , \ var\_3 \ term \ FROM \ var\_2 \end{array} \right\}$$

$$def\_plane \quad ::= \quad DEF \quad var\_1 \quad = \quad PLANE$$
$$\left\{ \begin{array}{l} point\_1 \ , \ point\_2 \ , \ point\_3 \\ point\_1 \ , \ var\_2 \\ var\_3 \\ term \ FROM \ var\_2 \\ EQUATION \ term \ , \ term \ , \ term \ = \ term \end{array} \right\}$$

$$point \quad ::=$$
$$\left\{ \begin{array}{l} (\ term \ , \ term < \ , \ term >< 0:1 >) \\ var \end{array} \right\}$$

It describes an object with the type *POINT*, *STRAIGHT* or *PLANE* and named with *var* in the following ways:

1. $DEF \ \ B = POINT(X,Y,Z)$: point with coordinates x, y, z
2. $DEF \ \ B = POINT \ \ INTERSECT \ \ G1, G2$: intersection between straightlines $G1$ and straightline or plane $G2$
3. $DEF \ \ B = POINT \ \ N \ \ FROM \ \ A$: N-th point of the object A
4. $DEF \ \ B = STRAIGHT \ \ P1, P2$: straightline between the points $P1$ and $P2$
5. $DEF \ \ B = STRAIGHT \ \ INTERSECT \ \ C, D$: intersection of the planes C and D
6. $DEF \ \ B = STRAIGHT \ \ N \ \ FROM \ \ S$: according to the N-th line of the polyline S
7. $DEF \ \ B = PLANE \ \ P1, P2, P3$: plane by three points
8. $DEF \ \ B = PLANE \ \ P, G$: plane by point P and line G
9. $DEF \ \ B = PLANE \ \ F$: plane containing the face F
10. $DEF \ \ B = PLANE \ \ N \ \ FROM \ \ A$: plane containing the N-th face of the object A
11. $DEF \ \ B = PLANE \ \ EQUATION \ \ A1, A2, A3 = Q$: plane with the equation $A1x + A2y + A3z = Q$

A point is defined by the coordinate triple (x, y, z) or by the pair (x, y) if $z = 0$, and by an already defined variable.

Examples:

$$
\begin{aligned}
DEF \ \ B &= POINT \ \ (N+K, 55, X);\\
DEF \ \ B &= POINT \ \ INTERSECT \ \ G1, G2;\\
DEF \ \ B &= POINT \ \ N+K+5 \ \ FROM \ \ A;\\
DEF \ \ B &= STRAIGHT \ \ INTERSECT \ \ E1, E2;\\
DEF \ \ B &= STRAIGHT \ \ N+K-5 \ \ FROM \ \ S;\\
DEF \ \ B &= PLANE \ \ P1, P2, P3;\\
DEF \ \ B &= PLANE \ \ P1, (1, N, N+3), (5, R, S);\\
DEF \ \ B &= PLANE \ \ P1, G1;\\
DEF \ \ B &= PLANE \ \ FL1;\\
DEF \ \ B &= PLANE \ \ V+K-5 \ \ FROM \ \ A;\\
DEF \ \ B &= PLANE \ \ EQUATION \ \ A1, A2, A3 = A4;
\end{aligned}
$$

Lines and Polylines

A straightline is considered to be a special case of a polyline. In defining a polyline, it is possible to hide some connections. Which of the single straightlines of a polyline are hidden can be exchanged by the command *INVERT*.

$$def\_pline \ ::=$$

$$DEF \ var\_1 \ = \left\{ \begin{array}{l} PLINE \ \ sign\_point\_list \\ LINE \ \ term \ \ FROM \ \ var\_2 \\ POLY \ \ object\_list \end{array} \right\}$$

$$invert \ ::= \ INVERT \ \ var$$

$$sign\_point \ ::= \ < \ + \ | \ - \ >< 0:1 > point$$

Semantics:

1. $DEF \ \ S \ = \ PLINE \ \ P1, ..., PN \ :$
   This is the definition of the polyline S by the points $P1, ..., PN$ which may carry a sign. A positive sign indicates that the next line is not hidden and a negative sign indicates that the next line is hidden.

2. $DEF \ \ S - LINE \ \ N \ \ FROM \ \ X :$
   S is defined as polyline (or only line) and represents the N-th edge of a face X and a polyline X, respectively.

3. $DEF \ \ S - POLY \ \ P1, P2, P3, P4 :$
   Definition of the shortest line that contains $P1$ until $P4$. If the points don't lie on a straight line, then a face of a solid arises (see below).

4. $INVERT \ \ S :$
   The hidden and non-hidden lines are exchanged in the polyline S.

Examples:

$DEF \ \ P = PLINE \ \ P1, P2, -P3, P4, (N+3, M*N, 6.0), P5, (1,1), P6;$
$DEF \ \ S = LINE \ \ N+K-5 \ \ FROM \ \ A;$

$INVERT \ \ Q;$

Elementary Planar Objects

$$def\_face \quad ::=$$

$$DEF \;\; var\_1 \;\; = \;\; \left\{ \begin{array}{l} RECTANGLE \;\; term \;\; , \;\; term \\ TRIANGLE \;\; term \;\; , \;\; term \;\; , \;\; term \\ CIRCLE \;\; term \\ POLY \;\; object\_list \\ FACE \;\; term \;\; FROM \;\; var\_2 \end{array} \right\}$$

Semantics:
Elementary planar objects are defined in the following way:

1. $DEF \;\; A = RECTANGLE \;\; L, B$:
   Rectangle with side $[0, L]$ of the x-axis.

2. $DEF \;\; A = TRIANGLE \;\; S1, S2, S3$:
   Triangle with side $[0, S1]$ of the x-axis, $S2$ and $S3$ are the lengths of the other sides written in the mathematically positive sense.

3. $DEF \;\; A = CIRCLE \;\; R$:
   Circle with radius R in the xy-plane.

4. $DEF \;\; A = POLY \;\; P1, P2, ..., PN$:
   Convex hull A of the elementary or connected objects $P1, P2, ..., PN$ belonging to a plane. If all points lie on a line, and not all points in a plane, then a straightline or a face arises (see below).

5. $DEF \;\; A = FACE \;\; N \;\; FROM \;\; X$:
   N-th surface of a body X.

Examples:

$$DEF \;\; B \;\; = \;\; RECTANGLE \;\; 2.5, X - H;$$
$$DEF \;\; B \;\; = \;\; TRIANGLE \;\; 4, 5, 6;$$
$$DEF \;\; B \;\; = \;\; CIRCLE \;\; 7.2;$$
$$DEF \;\; B \;\; = \;\; POLY \;\; F1, Q5, P5, (4, 3.2, 0), (3, N + K);$$
$$DEF \;\; B \;\; = \;\; FACE \;\; N + K - 5 \;\; FROM \;\; Q;$$

Elementary Spatial Objects

$$def\_body ::= DEF\ var\_1 = \left\{ \begin{array}{l} cuboid \\ pyramid \\ pyr\_stump \\ cylinder \\ cone \\ conestump \\ polyhedron \end{array} \right\}$$

$cuboid ::= CUBOID\ term\ ,\ term\ ,\ term$

$pyramid ::= PYRAMID\ point\ GROUNDF\ point\_list$

$pyr\_stump ::=$

$PSTUMP\ point\ HEIGHT\ term\ GROUNDF\ point\_list$

$cylinder ::= CYLINDER\ RADIUS\ term\ HEIGHT\ term$

$cone ::= CONE\ RADIUS\ term\ HEIGHT\ term$

$cone\_stump ::=$

$CSTUMP\ RADIUS\ term\ ,\ term\ HEIGHT\ term$

$polyhedron ::= POLY\ object\_list$

Semantics:

1. $CUBOID\ L, B, H :$
   The first side $(0, L)$ on the x-axis, the second side $(0, B)$ on the y-axis and the third side (0,H) on the z-axis.

2. $PYRAMID\ S\ GROUNDF\ P1, ..., PN :$
   Pyramid with peak S and base face.

3. $PSTUMP\ S\ HEIGHT\ H\ GROUNDF\ P1, ..., PN :$
   Truncated pyramid with the imaginary peak S.

4. $CYLINDER\ RADIUS\ R\ HEIGHT\ H :$
   Centrally positioned on the xy-plane.

5. $CONE\ RADIUS\ R\ HEIGHT\ H\ :$
   The cone is positioned like a cylinder.

6. $CSTUMP\ RADIUS\ R1, R2\ HEIGHT\ H :$
   Truncated cone, lower and upper radius.

7. Polyhedron as the convex hull over the (elementary or connected) objects.

Examples:

$$\begin{array}{lll} DEF\ A & = & CUBOID\ \ 2,3,A*B-1; \\ DEF\ B & = & PYRAMID\ \ (2,4,N)\ \ GROUNDF\ \ P,Q,R,(6,N+K); \\ DEF\ C & = & PSTUMP\ \ S1\ \ HEIGHT\ \ 4.5\ \ GROUNDF\ \ P,Q,R,S; \\ DEF\ D & = & CONE\ \ RADIUS\ \ N+K+3\ \ HEIGHT\ \ 2+7.44; \\ DEF\ E & = & CSTUMP\ \ RADIUS\ \ N+K-5,R\ \ HEIGHT\ \ 2+7.44; \\ DEF\ F & = & CYLINDER\ \ RADIUS\ \ N+7.44\ \ HEIGHT\ \ K; \\ DEF\ H & = & POLY\ \ P1,P2,(1,1,2.6),RO; \end{array}$$

Definition of Tree Objects by Special Operations

$$connection\ ::=\ TREE\ \ var\ \ =\ \ connection\_term$$

$$connection\_term\ ::=\ operand < \left\{ \begin{array}{c} + \\ - \\ * \end{array} \right\} operand >< 1 : >$$

$$operand\ ::=\ <\ (\ connection\_term\ )\ |\ var\ >$$

$$\begin{array}{l} boundary\ ::=\ BOUND\ \ var \\ bound\_back\ ::=\ BACK\ \ var \end{array}$$

Semantics:

1. A common object *var* is formed from partial objects by connections. By using the key word *TREE*, it is possible to connect objects of different dimensions such as bodies, faces, lines and points. A connection is made only by a common name. A new boundary representation does not arise. The positions relative to each other are not declared. It may be that they can touch, intersect, as well as be disjunctive to each other.

   $A+B$ means union (or only sum) of objects A and B.

   $A-B$ yields the difference of A and B, i.e., the part of A that does not belong to B.

   $A*B$ is the intersection, i.e., the part that belongs to A as well as to B.

The same priority rules as in arithmetic terms are valid: $'*'$ has the highest priority, $'+'$ and $'-'$ have the same priority, so that the computation goes from left to right.

2. *BOUND A* : The operation defining the common object are now computed. New vertices, edges and faces of the result are built. Only common objects are computed the single parts are all from the same dimension type.

3. *BACK A* : The boundary representation of an object is deleted again and an object arises that is only connected by name.

Examples:

$$TREE\ \ R\ =\ A - B * (C + D);$$
$$BOUND\ \ R;$$
$$BACK\ \ R;$$

Copy and Exchange of Objects

copy ::= *COPY var_1* = *var_2*
substitute ::= *SUBSTITUTE var_1* = *var_2*

Semantics:

1. The geometric object *var_1* is defined by copying the object *var_2*. Thereby, all of the connections and elementary objects attached to *var_2* are copied too. The modeler chooses new names for these objects. They are built by the first two letters of *var_2*, and a number chosen from the computer itself.

2. The previously defined object *var_1* can be exchanged by the object *var_2* as a part in all common objects connected by name. Both of the objects must be elementary ones.

Examples:

$$COPY\ \ V1 = V2;$$
$$SUBSTITUTE\ \ V1 = V2;$$

Resolving and Deleting of Objects

$$delete \ ::= \left\{ \begin{array}{l} DELETE \ \ var\_list \\ DELETE \ ALL \\ DISSOLVE \ \ var\_list \end{array} \right\}$$

Semantics:

1. $DELETE \ \ A1, ..., AN$: The objects $A1, ..., AN$ and their partial objects are deleted. The objects listed are traced and deleted in all tree objects.
2. $DELETE \ ALL$: All objects defined until this point are deleted.
3. $DISSOLVE \ \ A1, ..., AN$: The tree objects $A1, ..., AN$ are deleted, but not their partial objects.

Examples:

$$DELETE \ \ A, B, C;$$
$$DELETE \ \ ALL;$$
$$DISSOLVE \ \ A, B, C;$$

Computation of Vectors

$$vector\_computation \ ::= \ VECTOR \ \ var\_1 \ = \ vector\_term$$
$$vector\_term ::= vector\_term < \left\{ \begin{array}{c} + \\ - \end{array} \right\} vector\_term >< 0 : >$$
$$vector\_term \ \ ::= \ < \ scalar \ \ * \ >< 0 : 1 > name\_1$$
$$scalar \ \ ::= \ < \ name\_2 \ | \ integer \ >$$

Semantics:

var_1 is a vector. Vectors get the type $POINT$. They can be assumed to be a position vector of points in E^3. The syntax is as the following:

$A + B$	Addition of two vectors
$A - B$	Subtraction of two vectors
$K * A$	Multiplication of the vector A with the scalar factor K (result is real)

$name\_1$ is a name of a point and $name\_2$ is a name of a real or integer number.

Examples:

$$VECTOR \ \ V1 = 5 * V2 + P - K * Q;$$

Transformation of Objects

$$location \ ::= \left\{ \begin{array}{l} motion \\ rotation \\ orientation \end{array} \right\}$$
$motion \ ::= \ MOVE \ var \ VECTOR \ point$
$rotation ::= ROTATE \ var\_1 \ STRAIGHT \ var\_2 \ ANGLE \ arc$
$orientation \ ::= \ ORIENTATE \ var\_list$
$arc \ ::= \ term$

Semantics:

1. The object *var* is translated by the vector *POINT*.
2. The object *var_1* is rotated at the straightline *var_2* with an angle of the size *arc* by means of the right hand rule. The thumb points into the orientation of the line. The fingers wind themselves into the turn direction.
3. The orientation of the named objects (straightlines, faces, planes, etc.) can be exchanged.
4. If the angle is given only by a number then it is interpreted as an arc measure. If the letter D is added (without a blank) then the number is interpreted as a degree. Variables are always interpreted with respect to arc measure.

Examples:

$MOVE \ Z1 \ VECTOR \ (1, N + K, 5);$
$ROTATE \ Z2 \ STRAIGHT \ G1 \ ANGLE \ 1.541 * 2;$
$ROTATE \ Z2 \ STRAIGHT \ G1 \ ANGLE \ 10D;$
$MOVE \ Z1 \ VECTOR \ P;$
$ORIENTATE \ G1, EBE, F2;$

The Kind of Projection

$$projection ::= \left\{ \begin{array}{l} FRONTVIEW \mid SIDEVIEW \mid TOPVIEW \\ ISOMETRY \mid DIMETRY \mid CAVALIER \\ ORTHOGONAL < arc\_1 \quad , \quad arc\_2 >< 0:1 > \\ AXONOMETRY arc\_1 \ , \ arc\_2 \ < , \ el\_i < 3:3 > \end{array} \right\}$$

$$arc ::= term$$

$$el ::= term$$

Semantics:

Such a command chooses the kind of projection with the following parameters. The angles define the direction of the projection. The central projection takes a horizontal and vertical view sector of 30 degree.

Parameter		Semantic	Standard values
arc_1	:	expansion of the side view	45 (orthogonal)
arc_2	:	expansion of the vertical view	60 (axonometry)
el	:	elongations of the $x-, y-$ and z-axis	1.0, 1.0, 1.0

The angles are interpreted with respect to arc measure. If a D follows the number then it is interpreted as degrees. This character cannot follow a variable and cannot be interpreted as a variable.

Examples:

$$FRONTVIEW;$$
$$ISOMETRY;$$
$$CAVALIER;$$
$$ORTHOGONAL \quad A+B, 5D;$$
$$AXONOMETRY \quad A+BG, 5, DX, DY, DZ-3;$$
$$CENTRAL \quad A.1, A.2, A.3, P.1, P.2, P.3;$$

Hidden Lines and Approximation Lines

$$covering ::= COVERING \ : \left\{ \begin{array}{l} NO \\ HALF \\ COMPLETE \end{array} \right\}$$

$$approximation ::= APPROXIMATION \ : \left\{ \begin{array}{l} CONTOUR \\ ALL \end{array} \right\}$$

Semantics:

1.	*NO*	:	All of the lines are drawn.
	HALF	:	Lines and those faces pointing to the eye are represented.
	COMPLETE	:	Hidden lines are deleted (standard)

2. Either only the approximation lines of the contour or all of the approximation are drawn. The previously defined circle division parameter defines the segment angle to termine whether the approximation line is hidden or not. The default is *CONTOUR*.

Examples:

COVERING	:	*NO*;
COVERING	:	*HALF*;
COVERING	:	*COMPLETE*;
APPROXIMATION	:	*CONTOUR*;
APPROXIMATION	:	*ALL*;

Image Output

$$\begin{array}{lll} add & ::= & ADD\ \ var\_list \\ reduce & ::= & REDUCE \left\{ \begin{array}{l} var\_list \\ ALL \end{array} \right\} \\ picture & ::= & TO\ \ integer \end{array}$$

Semantics:

This actions *ADD* and *REDUCE* indicate how objects have to be prepared for a next common picture. The picture instruction describes the image output of those geometric objects that are written in the name storage of the image. Additionally it indicates the device to be used for drawing or the kind of image file representation. The logical graphics device expressions are assigned special importance:

1	graphic screen (VGA)	-
11	drawing file with the polylines in ASCII code	*.ASC
12	drawing file for use in AUTOCAD	*.DXF
13	drawing file for TEX	*.TEX
21	geometry with surface boundaries in ASCII code	*.FAC
22	geometry with convex patches in ASCII code	*.CON
23	geometry with vertex boundaries in ASCII code	*.VER

Examples:

$$\begin{array}{l} REDUCE \;\; ALL; \\ ADD \;\; A, B, C; \\ TO \;\; 1; \end{array}$$

Geometric Data Files

$$\begin{array}{lll} output & ::= & OUTPUT \;\; var \;\; EXT\,NAME \;\; name \;\; TO \;\; device \\ input & ::= & INPUT \;\; var \;\; EXT\,NAME \;\; name \;\; FROM \;\; device \\ device & ::= & integer \end{array}$$

Semantics:

The object *var* can be written to and be read from a data file with the name *name* and the type $'.Gn'$. The last two letters of the type correspond with *integer*.

Examples:

$$\begin{array}{l} OUTPUT \;\; Q \;\; EXT \;\; NAME \;\; BB \;\; TO \;\; 11; \\ INPUT \;\; Q \;\; EXT \;\; NAME \;\; BB \;\; FROM \;\; 11; \end{array}$$

Protocols

$$protocol ::= \left\{ \begin{array}{l} PROT \;\; integer \\ OBINF \;\; var \end{array} \right\}$$

Semantics:

1. The protocol information to the graphic terminal defines *integer* :

1	about the last error
2	object names of the model storage
3	parameter of the image
4	object names of the image
5	storage occupancy

2. The object information *OBINF* gives an alphanumeric description of any object *var*.

Examples:

$$\begin{array}{l} PROT \;\; 2; \\ OBINF \;\; A; \end{array}$$

The Condition and the DO-instruction

Conditional statements are necessary to describe programming branches, the DO-instructions to define a loop.

$$
\begin{array}{lll}
\mathit{condition} & ::= & \mathit{IF}\ \ \mathit{log\_term}\ \ \mathit{THEN} \\
 & & \mathit{commands\_1} \\
 & & <\mathit{ELSE}\ \ \mathit{commands\_2}>< 0:1 > \\
 & & \mathit{END} \\
 & & \\
\mathit{do\_loop} & ::= & \mathit{WHILE}\ \ \mathit{log\_term}\ \ \mathit{DO} \\
 & & \mathit{commands} \\
 & & \mathit{END}
\end{array}
$$

$$
\mathit{log\_term} ::= \mathit{log\_term} << \mathit{AND}\ \mid\ \mathit{OR} > \mathit{log\_term} >< 0 : >
$$

$$
\mathit{log\_term} ::=< \mathit{NOT} >< 0:1 > \left\{ \begin{array}{c} \mathit{term} \left\{ \begin{array}{c} EQ \\ NE \\ LT \\ GT \\ LE \\ GE \end{array} \right\} \mathit{term} \\ (\ \mathit{log\_term}\ ) \end{array} \right\}
$$

Semantics:

1. In the conditional statement, if *log_term* yields the value *TRUE* then the *commands_1* is taken, otherwise the *commands_2* is taken.

2. In the DO-instruction, if *log_term* is still *TRUE* then the *commands* is repeated again.

Abbreviations:	*EQ*	:	equal
	NE	:	not equal
	LT	:	less
	GT	:	greater
	LE	:	less or equal
	GE	:	greater or equal

   The priority rule of the logical operation is *NOT* before *AND* before *OR*.

3. The DO-instructions cannot compare with those of programming languages because it is slower.

Examples:

$$IF\ \ N\ \ GT\ \ 0\ \ AND\ \ N\ \ LT\ \ 5\ \ THEN\ \ B = A[\ I\ ];$$
$$N + 1;$$
$$ELSE\ \ A = C;$$

$$INT\ \ I = 0;$$
$$WHILE\ \ I\ \ LE\ \ N\ \ DO\ \ I = I + 1;$$
$$DEF\ \ A[\ I\ ] = CUBOID\ \ 1, 2 * I, I;$$
$$MOVE\ \ A[\ I\ ]\ \ VECTOR\ \ (1, 2 * I, I - 1);$$
$$END;$$

The Instruction Set and the Comment

$$commands\ ::=\ command\ \ <\ ;\ \ command\ ><0:>$$

$$command\ ::=$$

$$\left\{\begin{array}{l}
dialog \mid file \mid help \\
declaration \mid assignment \mid detail \\
def_point \mid def_straight \mid def_PLANE \\
def_pline \mid def_face \mid def_body \\
connection \mid boundary \mid bound_back \\
copy \mid substitute \mid delete \\
vector_computation \mid location \\
projection \mid covering \mid approximation \\
picture \mid add \mid reduce \\
output \mid input \mid protocol \\
condition \mid do_loop \\
comment
\end{array}\right\}$$

$$comment\ ::=\ /*\ \ <\ letter \mid digit\ ><0:>\ \ */$$

B

Coefficients of a Projected Curve

In Chapter 6 an implicit representation of a plane curve of degree four

$$\begin{aligned} x &= x(t) = t \\ y &= y(t) \quad \textit{with} \quad 0 = c_4 y^4 + c_3 y^3 + c_2 y^2 + c_1 y + c_0 \end{aligned}$$

was studied, which is defined as the projection of a space curve with height

$$z = z(t) = -\frac{(b_2(t) - a_2(t))}{(b_1(t) - a_1(t))}.$$

This curve is taken from an intersection of two surfaces of degree two:

$$\begin{aligned} 0 &= z^2 + a_1 z + a_2 \\ 0 &= z^2 + b_1 z + b_2 \end{aligned}$$

in Euclidean space E^3 with coefficients

$$\begin{aligned} a_1 &= a_1(x,y) = a_{11}x + a_{12}y + a_{13} \\ b_1 &= b_1(x,y) = b_{11}x + b_{12}y + b_{13} \\ a_2 &= a_2(x,y) = a_{21}x^2 + a_{22}y^2 + a_{23}xy + a_{24}x + a_{25}y + a_{26} \\ b_2 &= b_2(x,y) = b_{21}x^2 + b_{22}y^2 + b_{23}xy + b_{24}x + b_{25}y + b_{26}. \end{aligned}$$

If we take the curve parameter $t = x$, then the coefficients c_0, c_1, c_2, c_3 and c_4 of the implicit equation

$$0 = c_4 y^4 + c_3 y^3 + c_2 y^2 + c_1 y + c_0,$$

in the variable $y = y(t)$, define the desired implicit parametric representation.

$\mathbf{c_0} =$

$$a_{26}^2 - a_{13}a_{26}b_{13} + a_{26}b_{13}^2 + a_{13}^2b_{26} - 2a_{26}b_{26} -
a_{13}b_{13}b_{26} + b_{26}^2 +$$

$$(2a_{24}a_{26} - a_{13}a_{26}b_{11} - a_{13}a_{24}b_{13} - a_{11}a_{26}b_{13} + 2a_{26}b_{11}b_{13} +
a_{24}b_{13}^2 + a_{13}^2b_{24} - 2a_{26}b_{24} - a_{13}b_{13}b_{24} + 2a_{11}a_{13}b_{26} -
2a_{24}b_{26} - a_{13}b_{11}b_{26} - a_{11}b_{13}b_{26} + 2b_{24}b_{26})t +$$

$$(a_{24}^2 + 2a_{21}a_{26} - a_{13}a_{24}b_{11} - a_{11}a_{26}b_{11} + a_{26}b_{11}^2 - a_{13}a_{21}b_{13} -
a_{11}a_{24}b_{13} + 2a_{24}b_{11}b_{13} + a_{21}b_{13}^2 + a_{13}^2b_{21} - 2a_{26}b_{21} -
a_{13}b_{13}b_{21} + 2a_{11}a_{13}b_{24} - 2a_{24}b_{24} - a_{13}b_{11}b_{24} - a_{11}b_{13}b_{24} +
b_{24}^2 + a_{11}^2b_{26} - 2a_{21}b_{26} - a_{11}b_{11}b_{26} + 2b_{21}b_{26})t^2 +$$

$$(2a_{21}a_{24} - a_{13}a_{21}b_{11} - a_{11}a_{24}b_{11} + a_{24}b_{11}^2 - a_{11}a_{21}b_{13} +
2a_{21}b_{11}b_{13} + 2a_{11}a_{13}b_{21} - 2a_{24}b_{21} - a_{13}b_{11}b_{21} - a_{11}b_{13}b_{21} +
a_{11}^2b_{24} - 2a_{21}b_{24} - a_{11}b_{11}b_{24} + 2b_{21}b_{24})t^3 +$$

$$(a_{21}^2 - a_{11}a_{21}b_{11} + a_{21}b_{11}^2 + a_{11}^2b_{21} - 2a_{21}b_{21} -
a_{11}b_{11}b_{21} + b_{21}^2)t^4$$

$\mathbf{c_1} =$

$$2a_{25}a_{26} - a_{13}a_{26}b_{12} - a_{13}a_{25}b_{13} - a_{12}a_{26}b_{13} + 2a_{26}b_{12}b_{13} +
a_{25}b_{13}^2 + a_{13}^2b_{25} - 2a_{26}b_{25} - a_{13}b_{13}b_{25} + 2a_{12}a_{13}b_{26} -
2a_{25}b_{26} - a_{13}b_{12}b_{26} - a_{12}b_{13}b_{26} + 2b_{25}b_{26} +$$

$$(2a_{24}a_{25} + 2a_{23}a_{26} - a_{13}a_{25}b_{11} - a_{12}a_{26}b_{11} - a_{13}a_{24}b_{12} -
a_{11}a_{26}b_{12} + 2a_{26}b_{11}b_{12} - a_{13}a_{23}b_{13} - a_{12}a_{24}b_{13} -
a_{11}a_{25}b_{13} + 2a_{25}b_{11}b_{13} + 2a_{24}b_{12}b_{13} + a_{23}b_{13}^2 + a_{13}^2b_{23} -
2a_{26}b_{23} - a_{13}b_{13}b_{23} + 2a_{12}a_{13}b_{24} - 2a_{25}b_{24} - a_{13}b_{12}b_{24} -
a_{12}b_{13}b_{24} + 2a_{11}a_{13}b_{25} - 2a_{24}b_{25} - a_{13}b_{11}b_{25} - a_{11}b_{13}b_{25} +
2b_{24}b_{25} + 2a_{11}a_{12}b_{26} - 2a_{23}b_{26} - a_{12}b_{11}b_{26} - a_{11}b_{12}b_{26} +
2b_{23}b_{26})t +$$

$$(2a_{23}a_{24} + 2a_{21}a_{25} - a_{13}a_{23}b_{11} - a_{12}a_{24}b_{11} -
a_{11}a_{25}b_{11} + a_{25}b_{11}^2 - a_{13}a_{21}b_{12} - a_{11}a_{24}b_{12} + 2a_{24}b_{11}b_{12} -
a_{12}a_{21}b_{13} - a_{11}a_{23}b_{13} + 2a_{23}b_{11}b_{13} + 2a_{21}b_{12}b_{13} +
2a_{12}a_{13}b_{21} - 2a_{25}b_{21} - a_{13}b_{12}b_{21} - a_{12}b_{13}b_{21} +
2a_{11}a_{13}b_{23} - 2a_{24}b_{23} - a_{13}b_{11}b_{23} - a_{11}b_{13}b_{23} +
2a_{11}a_{12}b_{24} - 2a_{23}b_{24} - a_{12}b_{11}b_{24} - a_{11}b_{12}b_{24} + 2b_{23}b_{24} +
a_{11}^2b_{25} - 2a_{21}b_{25} - a_{11}b_{11}b_{25} + 2b_{21}b_{25})t^2 +$$

$$(2a_{21}a_{23} - a_{12}a_{21}b_{11} - a_{11}a_{23}b_{11} + a_{23}b_{11}^2 - a_{11}a_{21}b_{12}+$$
$$2a_{21}b_{11}b_{12} + 2a_{11}a_{12}b_{21} - 2a_{23}b_{21} - a_{12}b_{11}b_{21}-$$
$$a_{11}b_{12}b_{21} + a_{11}^2b_{23} - 2a_{21}b_{23} - a_{11}b_{11}b_{23} + 2b_{21}b_{23})t^3$$

$\mathbf{c_2} =$

$$a_{25}^2 + 2a_{22}a_{26} - a_{13}a_{25}b_{12} - a_{12}a_{26}b_{12} + a_{26}b_{12}^2-$$
$$a_{13}a_{22}b_{13} - a_{12}a_{25}b_{13} + 2a_{25}b_{12}b_{13} + a_{22}b_{13}^2 + a_{13}^2b_{22}-$$
$$2a_{26}b_{22} - a_{13}b_{13}b_{22} + 2a_{12}a_{13}b_{25} - 2a_{25}b_{25} - a_{13}b_{12}b_{25}-$$
$$a_{12}b_{13}b_{25} + b_{25}^2 + a_{12}^2b_{26} - 2a_{22}b_{26} - a_{12}b_{12}b_{26} + 2b_{22}b_{26}+$$

$$(2a_{22}a_{24} + 2a_{23}a_{25} - a_{13}a_{22}b_{11} - a_{12}a_{25}b_{11} - a_{13}a_{23}b_{12}-$$
$$a_{12}a_{24}b_{12} - a_{11}a_{25}b_{12} + 2a_{25}b_{11}b_{12} + a_{24}b_{12}^2 - a_{11}a_{22}b_{13}-$$
$$a_{12}a_{23}b_{13} + 2a_{22}b_{11}b_{13} + 2a_{23}b_{12}b_{13} + 2a_{11}a_{13}b_{22}-$$
$$2a_{24}b_{22} - a_{13}b_{11}b_{22} - a_{11}b_{13}b_{22} + 2a_{12}a_{13}b_{23} - 2a_{25}b_{23}-$$
$$a_{13}b_{12}b_{23} - a_{12}b_{13}b_{23} + a_{12}^2b_{24} - 2a_{22}b_{24} - a_{12}b_{12}b_{24}+$$
$$2b_{22}b_{24} + 2a_{11}a_{12}b_{25} - 2a_{23}b_{25} - a_{12}b_{11}b_{25} - a_{11}b_{12}b_{25}+$$
$$2b_{23}b_{25})t+$$

$$(2a_{21}a_{22} + a_{23}^2 - a_{11}a_{22}b_{11} - a_{12}a_{23}b_{11}+$$
$$a_{22}b_{11}^2 - a_{12}a_{21}b_{12} - a_{11}a_{23}b_{12} + 2a_{23}b_{11}b_{12} + a_{21}b_{12}^2+$$
$$a_{12}^2b_{21} - 2a_{22}b_{21} - a_{12}b_{12}b_{21} + a_{11}^2b_{22} - 2a_{21}b_{22}-$$
$$a_{11}b_{11}b_{22} + 2b_{21}b_{22} + 2a_{11}a_{12}b_{23} - 2a_{23}b_{23} - a_{12}b_{11}b_{23}-$$
$$a_{11}b_{12}b_{23} + b_{23}^2)t^2$$

$\mathbf{c_3} =$

$$2a_{22}a_{25} - a_{13}a_{22}b_{12} - a_{12}a_{25}b_{12} + a_{25}b_{12}^2 - a_{12}a_{22}b_{13}+$$
$$2a_{22}b_{12}b_{13} + 2a_{12}a_{13}b_{22} - 2a_{25}b_{22} - a_{13}b_{12}b_{22} - a_{12}b_{13}b_{22}+$$
$$a_{12}^2b_{25} - 2a_{22}b_{25} - a_{12}b_{12}b_{25} + 2b_{22}b_{25}+$$

$$(2a_{22}a_{23} - a_{12}a_{22}b_{11} - a_{11}a_{22}b_{12} - a_{12}a_{23}b_{12} + 2a_{22}b_{11}b_{12}+$$
$$a_{23}b_{12}^2 + 2a_{11}a_{12}b_{22} - 2a_{23}b_{22} - a_{12}b_{11}b_{22} - a_{11}b_{12}b_{22}+$$
$$a_{12}^2b_{23} - 2a_{22}b_{23} - a_{12}b_{12}b_{23} + 2b_{22}b_{23})t$$

$\mathbf{c_4} =$

$$a_{22}^2 - a_{12}a_{22}b_{12} + a_{22}b_{12}^2 + a_{12}^2b_{22} - 2a_{22}b_{22} - a_{12}b_{12}b_{22} + b_{22}^2$$

Bibliography

[Ab 91] Abramowsky, S.; Müller, H.: Geometrisches Modellieren. Reihe Informatik, Band 75, BI Wissenschaftsverlag Mannheim, Wien, Zürich, 1991.

[Al 90] Alagar, V.S.; Bui, T.D.; Periyasamy, K.: Semantic CSG trees for finite element analysis. CAD, 22(4)1990.

[Ar 90] Arbab,F.: Set models and boolean operations for solids and assemblies. Computer Graphics and Applications, IEEE Computer Society, 9(1990)76-86.

[Ba 88] Bajaj, C.L.; Hoffmann, C.M.; Lynch, R.E.; Hopcroft, J.E.H.: Tracing surface intersections. CAGD, 5(4)1988.

[Ba 90] Barsalou, T.; Wiederhold, G.: Complex objects for relational databases. CAD, 22(8)1990.

[Be 90] Bez, H.E.; Edwards, J.: Distributed algorithm for the planar convex hull problem. CAD, 22(2)1990.

[Bi 90] Bieri, H.; Schmidt, P.M.: An on-line algorithm for constructing sweep planes in regular position. Preprint IAM-90-015, Universität Bern, 1990.

[Bl 88] Bloomenthal, J.: Polygonalization of implicit surface. CAGD, 5(4)1988.

[Bl 89] Bloor, M.I.G.; Wilson, M.J.: Generating blend surfaces using partial differential equations. CAD, 23(3)1989.

[Bo 90] de Boor, C.: Local corner cutting and the smoothness of the limiting curve. CAGD, 7(5)1990.

[Br 89] Brändli, N.; Mittelstaedt, M.: Exchange of solid models: current state and future trends. CAD, 21(2)1989.

[Br 88] Bronsvoort, W.F.: Boundary evaluation and direct display of CSG models. CAD, 20(7)1988.

[Br 89] Bronsvoort, W.F.; Garnaat, H.: Incremental display of CSG models using local updating. CAD, 21(4)1989.

[Ca 88] Callahan, M.; Hoffmann, D.; Hoffmann,T.: Computer graphics tools for the study of minimal surfaces. Communications of the ACM, Volume 31/6, 1988.

[Ch 89] Choi, B.K.; Ju, S.Y.: Constant-radius blending in surface modelling. CAD, 21(4)1989.

[Ch 90] Choi, B.K.; Lee, C.S.: Sweep surfaces modelling via coordinate transformations and blending. CAD, 22(2)1990.

[Da 91] Day, A.M.: Parallel implementation of 3D convex-hull algorithm. CAD, 23(3)1991.

[Do 91] Dobkin, D.P.; Souvaine, D.L.: Detecting the intersection of convex objects in the plane. CAGD, 8(3)1991.

[Du 88] Duong Kieu Hoa: Ein allgemeines Prinzip zur Modellierung von Objekten im dreidimensionalen Raum. Dissertation A, TU Dresden, 1988.

[En 88] Encarnacao, J.; Straßer, W.: Computer Graphics. R. Oldenbourg-Verlag München, Wien, 1988.

[Fa 88] Farin, G.: Curves and surfaces for computer-aided geometric design. Academic Press Boston, 1988.

[Fa 91] Farouki, R.T.; On the stability of transformations between power and Bernstein polynomial forms. CAGD, 8(1991)29-36.

[Fa 90] Farouki, R.T.; Neff, C.A.: Algebraic properties of plane offset curves. CAGD, 7(1990)101-127.

[Fa 88] Farouki, R.T.; Rajan, V.T.: On the numerical condition of algebraic curves and surfaces 1. Implicit equations. CAGD, 5(8)1988.

[Fe 90] Ferreira, J.C.E.; Hinduja, S.: Convex hull-based feature-recognition method for 2.5D components. CAD, 22(1)1990.

[Fi 91] Fischer, A.; Shpitalni, M.: Accelerating the evaluation of volumetric modelers by manipulating CSG trees and DAGS. CAD, 23(5)1991.

[Fi 90] Fitzhorn, P.A.: Language of topologically valid bounding manifolds. CAD, 22(7)1990.

[Ga 91] Garrity, T.; Warren, J.: Geometric continuity. CAGD, 8(1)1991.

[Gu 89] Guibas, L.; Salesin, D.; Stolfi, J.: Epsilon geometry: Building robust algorithms from imprecise computations. ACM Symp. on Computational Geometry, (1989)208-217.

[Gu 88] Günther, O.: Efficient structures for geometric data management. Lecture Notes in Computer Science, Springer-Verlag New York, London, Paris, Tokyo, 1988.

[Gu 89] Günther, O.; Wong, E.: Convex polyhedral chains: a representation for geometric data. CAD, 21(3)1989.

[Gu 91] Gursoz, E.L.; Choi, Y.; Prinz, F.B.: Boolean set operations on non-manifold boundary representation objects. CAD, 23(1)1991.

[Ha 90] Hansford, D.: The neutral case for the min-max triangulation. CAGD, 7(5)1990.

[Ha 90] Hartmann, E.: Blending of implicit surfaces with functional splines. CAD, 22(8)1990.

[Ha 87] Hartwig, A.; Mager, K.: 3D-Modellierung mit GEKO. WBZ MKR/IV der TU Dresden, Studientexte Computergeometrie, 86(1986).

[Ha 88] Hartwig, A.: GEKO 1800, Anwendungsbeschreibung. ROBOTRON-Vertrieb Dresden, 1988.

[Ha 89] Zur Stabilität von Algorithmen der 3D-Modellierung. DIGRA 88 - Computergraphik und Anwendung, Rostocker Informatikberichte, Heft 8/1989.

[Ho 87] Hoffmann, C.; Hopcroft, J.; Karasick, M.: Robust set operation on poyhedral solids. Tech. Rept. 723, Comp. Sci., Purdue University, 1987.

[Ho 87] Hoffmann, C.M.; Hopcroft, J.E.: Geometric ambiguities in boundary representations. CAD, 19(3)1987.

[Ho 88] Hoffmann, C.; Hopcroft, J.; Karasick, M.: Towards implementing robust geometric computations. ACM Symp. on Computational Geometry, (1988)106-117.

[Ho 89] Hoffmann, Ch.M.: Geometric and solid modeling. Morgan Kaufmann Publishers San Mateo, 1989.

[Ho 90] Hoffmann, C.M.: A dimensionality paradigm for surface interrogations. CAGD, 7(6)1990.

[Hl 88] Holmström, L; Laakko, T.: Rounding facility for solid modelling of mechanical parts. CAD, 20(10)1988.

[Ho 89] Hoschek, J.; Lasser, D.: Grundlagen der geometrischen Datenverarbeitung. B.G. Teubner Stuttgart, 1989.

[Ka 89] Kalay, Y.E.: The hybrid edge: a topological data structure for vertically integrated geometric modelling. CAD, 21(3)1989.

[Ki 90] Kim, S.H.; Lee, K.: An assembly modelling system for dynamic and kinematic analysis. CAD, 21(1)1989.

[Ki 90] Kim, W.; Banerjee, J.; Chou, H.-T., Garza, J.F.: Object-oriented database support for CAD. CAD, 22(8)1990.

[Kn 78] Knuth, D.E.: The art of computer programming. Addison Wesley Reading, 1978.

[Ko 87] Kohn, M.: Practical numerical methods - algorithms and programs. McGraw-Hill New York, 1987.

[Ko 91] Kosters, M.: High-order implicit blending surfaces of low degree. CAGD, 8(5)1991.

[Kr 86] Kreyszig, E.: Differentialgeometrie. Akademische Verlagsgesellschaft Leipzig, 1968.

[Kr 90] Kriezis, G.A.; Prakash, P.V.; Patrikalakis, N.M.: A method for intersecting algebraic surfaces with rational polynomials patches is presented. CAD, 22(10)1990.

[Ku 68] Kuratowski, K.; Mostowski, A.: Set theory. Studies in logic and the foundations of mathematics, North-Holland Publishing Company Amsterdam, 1968.

[Ku 88] Kutzler, B.: Algebraic approaches to automated geometry proving, Ph.D.Diss., Res. Institute for Symb. Comp., Rept. 88-74.0, Kepler University Linz, 1988.

[Ku 89] Kulisch, U.; Geörg, S.; Hammer, R.; Ralz, D.: Wissenschaftliches Rechnen mit Ergebnisverifikation. Akademieverlag Berlin, 1989.

[La 86] Lancaster, P.; Salkauskas, K.: Curve and surface fitting. Academic Press London, 1986.

[Le 79] Levin, J.: Mathematical models for determining the intersections of quadric surfaces. Comp. Vision, Graphics and Image Processing, 11(1979)73-87.

[Li 90] Li, Z.; Milenkovic, V.: Constructing strongly convex hulls using exact or rounded arithmetic. ACM Symp. on Computational Geometry, (1990)235-243.

[Lo 90] Lowe, T.W.; Bloor, M.I.G.; Wilson, M.J.: Functionality in blend design. CAD, 22(10)1990.

[Ma 85] Maess, G.: Vorlesungen über numerische Mathematik. Birkhäuser Verlag Basel, Boston, Stuttgart, 1985.

[Ma 88] Mäntylä, M.: An introduction to solid modeling. Computer Science Press Rockville, 1988.

[Ma 90] Maripuri, S.R.; Zeid, I.: Generating aspect graphs for nonconvex polyhedra. CAD, 22(5)1990.

[Ma 89] Markot, R.P.; Magedson, R.L.: Solutions of tangential surface and curve intersections. CAD, 21(7)1989.

[Ma 88] Martin, R.R.; Stephenson, P.C.: Putting objects into boxes. CAD, 20(9)1988.

[Ma 90] Martin, R.R.; Stephenson, P.C.: Sweeping of three-dimensional objects. CAD, 22(4)1990.

[Mo 91] de Montoudouin, Y.: Resolution of $P(x, y) = 0$. CAD, 23(9)1991.

[Mu 90] Müllenheim, G.: Convergence of a surface/surface intersection algorithm. CAGD, 7(5)1990.

[Mu 91] Müllenheim, G.: On determining start points for a surface/surface intersection algorithm. CAGD, 8(5)1991.

[Ne 90] Nef, W.; Schmidt, P.M.: Computing a sweepingplane in regular ("general") position: a numerical and a symbolic solution. J. Symbolic Computation, 10(1990)633-646.

[Ne 89] Neff, C.A.: Decomposing algebraic sets using Gröbner bases. CAGD, 6(3)1989.

[Ne 83] Nestler, Th.: Zur Modellierung geometrischer Objekte. Forschungsbericht, TU Dresden, 1983.

[Pa 89] Paoluzzi, A.; Ramella, M.; Santarelli, A.: Boolean algebra over linear polyhedra. CAD, 21(8)1989.

[Pi 85] Piegl, L.: Representation of quadric primitives by rational polynomials. CAGD, 2(1-3)1985.

[Pi 89] Piegl, L.: Geometric method of intersecting natural quadrics represented in trimmed surface form. CAD 21(4)1989.

[Po 90] Pottmann, H.: Smooth curves under Tension. CAD, 22(4)1990.

[Po 81] Powell, M.J.D.: Approximation theory and methods. Cambridge University Press Cambridge, London, New York, New Rochelle, Melbourne, Sydney, 1981.

[Pr 85] Preparata, F.P.; Shamos, M.I.: Computational geometry. Springer-Verlag New York, 1985.

[Pr 90] Preparata, F.P.: Cyclides in computer aided geometric design. CAGD, 7(1990)221-242.

[Qu 79] Querenburg, B.v.: Mengentheoretische Topologie. Springer-Verlag Berlin, Heidelberg, New York, 1979.

[Re 80] Requicha, A.A.G.: Representations of rigid solid objects. Lecture Notes in Computer Science, 89(1980).

[Ro 87] Rocheleau, D.N.; Lee, K.: System for interactive assembly modelling. CAD, 19(2)1987.

[Ro 91] Rossignac, J.R.; Requicha, A.A.: Constructive Non-Regularized Geometry. CAD, 23(1)1991.

[Ro 91] Roy, U.; Liu, C.R.; Woo, T.C.: Review of dimensioning and tolerancing: representation and processing. CAD, 23(7)1991.

[Sa 88] Saeed, S.E.O.; de Pennington A., Dodsworth, J.R.: Offsetting in geometric modelling. CAD, 20(2)1988.

[Sa 91] Sapidis, N.; Perucchio, R.: Delaunay triangulation of arbitrarily shaped planar domains. CAGD, 8(6)1991.

[Sa 83] Sarraga, R.F.: Algebraic methods for intersections of quadratic surfaces, in: GM Solid. Comp. Graphics Image Proc. 22(1983)222-238.

[Sc 91] Schmidt, P.M.: About the precision of the coefficients of a sweeping-plane. Geometrical Problems of Image Processing, Proc. of the 5th Workshop in Georgenthal, (1981)107-113.

[Se 90] Segal, M.: Using tolerances to guarantee valid polyhedral modeling results. Computer Graphics, 24(4)1990.

[Se 85] Segal, M.; Sequin, C.: Consistent calculations for solids modeling. ASM Symp. on Computational Geometry, (1985)29-38.

[Se 89] Seidel, H.-P.: Polynome, Splines und symmetrische rekursive Algorithmen im Computer Aided Geometric Design. Habilitationsschrift, Universität Tübingen 1989.

[Se 52] Semple, J.G.; Kneebone, G.T.: Algebraic Projective Geometry. Clarendon Press Oxford, 1952.

[Se 59] Semple, J.G.; Kneebone, G.T.: Algebraic Curves. Clarendon Press Oxford, 1959.

[Se 85] Sederberg, T.W.: Piecewise Algebraic Surface Patches. Comp. Aided Geom. Design, 2(1985)53-59.

[Se 86] Sederberg, T.W.: Algebraic Geometry for Computer-Aided Geometric Design. IEEE Comp. Graphics and Applications, 6(1986)52-59.

[Se 89] Sederberg, T.W.: Algorithm for algebraic curve intersection. CAD, 21(9)1989.

[Sh 91] Shapiro,V.; Vossler, D.L.: Construction and optimization of CSG representations. CAD, 23(1)1991.

[Si 91] Sittas, E.: 3D desingn regerence framework. CAD, 23(5)1991.

[Sm 90] Smith, N.S.: Spatial data models and data strictures. CAD, 22(3)1990.

[Sp 73] Späth, H.: Algorithmen für elementare Ausgleichsmodelle. R. Oldenbourg-Verlag München, Wien, 1973.

[Sp 74] Späth, H.: Algorithem für multivariable Ausgleichsmodelle. R. Oldenbourg-Verlag München, Wien, 1974.

[St 89] Straßer, W.; Seidel, H.-P.: Theory and practice of geometric modeling. Springer-Verlag Berlin, Heidelberg, New York, London Paris, Tokyo, Hong Kong, 1989.

[Sz 91] Szilvasi-Nagy, M.: Flexible rounding operation for polyhedra. CAD, 23(9)1991.

[Oc 89] O'Conner, M.A.: Natural quadrics: Projections and intersections. IBM J. Res. Develop., 33(4)1989.

[On 87] Onodera, T.; Kawai, S.: A formal model of visualization in computer graphics systems. Lecture Notes in Computer Science, Springer-Verlag Berlin, Heidelberg, New York, London, Paris, Tokyo 1987.

[Ov 90] van Overveld, C.W.A.M.: Discrete bilinear blending and its application in rendering curved surfaces. CAD, 22(6)1990.

[To 63] Tood, J.: Introduction to the constructive theory of functions. Birkhäuser Verlag Basel, Stuttgart, 1963.

[Ta 91] Tang, K.; Woo, T.: Algorithmic aspects of alternating sums of volumes. Part 1: Data structure and difference operation. CAD, 23(5)1991.

[Tu 87] Turner, J.U.: Tolerances in computer-aided geometric design. Ph.D. Dissertation, Rensselaer Polytechnic Institute, 1987.

[Tu 88] Turner, J.U.: Accurate solid modeling using polyhedral approximations. IEEE Computer Graphics and Applications, 1988.

[Wa 50] Walker, R.: Algebraic curves. Springer-Verlag New York, 1950.

[Vo 87] Vogel, V.: Polyederalgebren in der Computergeometrie. Dissertation A, TU Dresden, 1987.

[Wa 88] Waggenspack, J.; Anderson W.N. and D.C.: Piecewise parametric approximations for algebraic curves. CAGD, 6(1)1989.

[Wa 39] Von der Warden, B.L.: Einführung in die algebraische Geometrie. Springer-Verlag Berlin, 1939.

[Wh 04] Whitesitt, J.E.: Boolesche Algebra und ihre Anwendungen. Vieweg-Verlag Braunschweig, 1964.

[Wi 90] Wilf, I.; Manor, Y.: Projective splitting of quadric faces. CAD, 22(8)1990.

[Ya 88] Yap, C.K.: A Geometric consistency theorem for a symbolic perturbation theorem. ACM Symp. on Computational Geometry, (1988)134-142.

[Zo 89] Zhou, X.: Parametrisierung der Flächen zweiten Grades und Konvertierung in die NURBS-Tensorproduktdarstellung. Forschungsbericht WSI-89-1, Universität Tübingen, 1989.

Index